Godfrey Sturgeon

Seventy Years of Farm Tractors

Seventy Years of Farm Tractors
1930 – 2000

Brian Bell MBE

OLD POND PUBLISHING

IPSWICH

First published 2012

ISBN 978-1-908397-34-8

A catalogue record for this book is available from the British Library

Published by
Old Pond Publishing Ltd
Dencora Business Centre
36 White House Road
Ipswich IP1 5LT
United Kingdom

www.oldpond.com

Cover design and book layout by Liz Whatling
Printed in China

Contents

Godfrey Sturgeon
Bungay Home.

Conversion Table

The following information is offered to help readers who may be too young to remember the imperial days of pounds, shilling and pence, hundredweights, gallons and yards.

1 gallon = 4.6 litres

1 inch = 25.4 mm

1 foot = 300 mm

1 yard = 910 mm

1 acre = 0.4 hectare

1cwt = 50.9 kg

1 ton = 1,016 kg

1 horsepower = 746 watts

1 shilling = 5 pence

£1 = 20 shillings

Introduction

By referencing over one hundred different makes of farm tractor, *Seventy Years of Farm Tractors* traces the evolution of the tractor from the starting handles and pan seats of the 1930s to the 21st century four-wheel drive tractor with an air-conditioned cab and a computer-management system. There were about 30,000 tractors on British farms in the early 1930s and this number had risen to just over 261,000 by 1948. Although there were considerably fewer farms and farm workers in 2000 compared with the 1930s the tractor population had grown to well in excess of 500,000 and most farms had more tractors than people to drive them.

Harry Ferguson's hydraulic system sparked a revolution in power farming at the end of World War Two but even then many of the tractors made in the 1920s and 1930s were still giving sterling service. It was not unusual in the late 1940s to see a threshing contractor hauling his drum and elevator from farm to farm with a steam traction engine. Horses were still a major source of farm power and tractors were little more than mechanical horses with a drawbar to pull trailed implements, a belt pulley to drive a stationary baler or chaff cutter and a power take-off shaft to drive a binder or a potato digger. The average power of farm tractors in the late 1940s, when there were still many holdings of less than 20 acres, was only 25 hp. Small ride-on tractors with 10 hp or less under the bonnet were replacing the horse on some holdings and their contribution to the mechanisation of small farms has been recognised in this book.

The UK tractor population had passed the 360,000 mark in 1960 and although most of the best-known names in the tractor world were still in business some were destined to disappear within the next twenty years. Average tractor power was approaching 65 hp in 1960 when models in the 45–95 hp bracket accounted for more than 90 per cent of total sales. Four-wheel drive tractors were becoming popular and with most manufacturers building their own four-wheel drive models by the late 1970s the earlier activities of County Commercial Cars when Roadless Traction came to an end.

More than half of the new tractors sold to British farmers in 1985 had four-wheel drive with a similar percentage in the 80–100 hp bracket. The change to four-wheel drive increased at a fast pace and by the early 1990s 85 per cent of new tractors over 40 hp had four-wheel drive. Almost 20,000 new tractors were registered for the first time in 1997 when the average tractor power was just over 110 hp. By the turn of the century in-cab computers were monitoring and controlling the engine and transmission, managing headland turns, measuring the area drilled or sprayed and even diagnosing engine faults.

From time to time farming magazines provide their readers with basic details and prices of tractors and farm machinery. A magazine published in 1958 listed twenty-six different manufacturers and eighty-six models of wheeled and crawler tractors but fifty years later there were eighteen different manufacturers and 290 models of wheeled and crawler tractors over 50 hp on the British market.

Where registered trade names are given they have only been used to identify a particular system of machine. Dates and periods of production are mainly taken from manufacturers' literature and even here conflicting information can be found. In some circumstances tractors were stockpiled and sold for considerable periods after production ceased. In other cases new models were introduced to the farming public many months before they went into quantity production.

Most of the photographs in this book are from my own collection or from early sales literature but I am also grateful to Stuart Gibbard for reading the manuscript and for the loan of photographs, which I have acknowledged in the captions.

Brian Bell
Suffolk, 2012

Chapter I

Early Days

The first petrol-engined motor car was built in Germany in 1885 and it is generally accepted that a tractor with an internal combustion engine was made in America in 1889. The British-made Hornsby-Akroyd oil-engined tractor did not appear until 1896. Early tractors, especially those of North American origin, looked like and weighed almost as much as a steam engine. They were well suited to belt work but most were generally too heavy for fieldwork.

British engineers, including Dan Albone and HP Saunderson, had developed smaller and more manoeuvrable oil-engined tractors at the beginning of the twentieth century. Dan Albone, who owned a bicycle repair business at Biggleswade, built a lighter and more compact tractor that might one day replace the horse.

The first tricycle-wheeled Ivel was demonstrated to farmers in 1902. Albone formed Ivel Agricultural Motors and the first Ivel-built tractor was sold in 1903. It had a forward and reverse gear and the driver's seat was conveniently placed, at least in the winter, by the side of the 30 gallon water tank used to cool the 22 hp twin-cylinder horizontally opposed petrol engine. The 1908 Saunderson Model F, also made at Bedford, had a twin-cylinder paraffin engine and a three forward and one reverse gearbox with a top speed of 6 mph. Marshalls of Gainsborough, Albone with his Ivel, and Saunderson were some of the more important British tractor makers in the early years of the twentieth century when Britain was the world's leading tractor exporter. The first Marshall tractor had a 30–35 hp twin-cylinder paraffin

1. The Ivel with a 22 hp twin-cylinder petrol engine was named after a river in Bedfordshire.

9

engine and a three forward and one reverse gearbox with a top speed of 6 mph. Later models were known as the Colonial.

The history of Case, Caterpillar, John Deere, International Harvester, Massey-Harris and other North American tractor pioneers dates back to the 1890s and early 1900s. International Harvester, formed when a group of harvest machinery manufacturers merged in 1902, made their first friction drive Gasoline Traction Engines in 1906. Other International Harvester tractors, many of them partly disguised as traction engines, appeared before the 4½ ton Mogul 12-25, which was launched in 1912. It had a twin-cylinder radiator-cooled horizontally opposed engine that started on petrol and switched to paraffin when it was hot, a two forward speed and one reverse gearbox with a top speed of 4 mph and the luxury of a cab. The popular one forward and one reverse gear Mogul 8-16 with single-cylinder hopper-cooled engine, suitable for small farms, appeared in 1914 and the twin-cylinder Titan 10-20 with two forward gears and one reverse was added in 1915. Both tractors had a chain drive to the rear wheels and a belt pulley. The single-cylinder

2. The seat and steering wheel were offset on the International Harvester Titan.

3. The radiator was behind the engine on the International Junior 8-16, which weighed a relatively light 1½ tons.

4. The Walsh & Clark Victoria ploughing engine with a twin-cylinder horizontally opposed engine appeared in 1915. The fuel tank, shaped like a steam engine boiler, held enough paraffin for four days work.

Mogul 10-20, also with two forward gears and one reverse and roller chain final drive, was added in 1916. The Mogul had a muslin cloth over the engine air intake to catch dust and the Titan had a sheep's wool dust catcher. Mogul tractors were made in Chicago and Titans in Milwaukee.

McCormick dealers sold the Mogul to American farmers and Deering dealers marketed the Titan. The first Mogul arrived in the UK in 1915 followed by the Titan in 1916 and some were included in the 6,000 or so tractors imported by the British Ministry of Munitions to help boost food production during World War One. More powerful models were added but when the more modern-looking International 8-16 was launched in 1917 time was running out for the Mogul and the Titan. The Chicago-built Junior 8-16 with a four-cylinder overhead-valve petrol or paraffin engine had three forward gears and one reverse, a multi-plate clutch, a band brake, an exposed roller chain drive to the rear axle and coil spring suspension on the front axle.

Cable ploughing with a steam engine at each side of a field using a balance plough was common practice in the early 1900s. With the growing popularity of the internal combustion engine John Fowler introduced a paraffin-engined tractor for cable ploughing in 1911 and the Walsh and Clark ploughing engine appeared in 1912. Both looked like steam engines with a cable-winding drum, presumably in an attempt to attract farmers still using steam power. The cable drum on the Walsh and Clarke Victoria held about 450 yards of 1½ in diameter steel rope and a pair of these engines ploughed 7–10 acres in a day.

Farm tractors in the early years of the twentieth century were little more than mechanical horses used to pull implements and their steel wheels limited the top speed to a few miles per hour. A belt pulley, used on steam engines to drive threshing machines and other stationary farmyard equipment, was standard equipment on most tractors by the early 1900s. In 1918 the International Harvester Junior 8-16 was one of the first tractors to have a power take-off shaft. A mechanical power lift appeared in 1927 and the Caterpillar 60 crawler, introduced in 1931, was the first tractor with a full diesel engine. Low-pressure pneumatic tyres replaced the steel wheels on an Allis-Chalmers Model U in 1932 and the Minneapolis-Moline UDLX Comfort

5. A Hercules engine was used for the Fordson Model F tractor until 1918 when it was replaced with an engine designed and built by the Ford Motor Co.

Tractor of 1938 had the added luxury of an enclosed cab with a radio, heater and windscreen wiper.

Fordson tractors date back to 1905 when Henry Ford made an experimental tractor based on a Model B car. Several more tractors, also developed from Ford cars, including the Eros conversion of a Model T car, appeared over the next ten years. Some had a Henry Ford & Son nameplate although later models were badged as Fordsons. The first Fordson Model F tractors were made at Dearborn in Michigan between 1917 and 1920 when production was transferred the short distance to Rouge River. The Model F, also made at Cork between 1919 and 1922, was the world's first mass-produced tractor. The Ministry of Munitions imported about 6,000 tractors with four-cylinder Hercules engines, made by Ford at Dearborn from 1917–18, to help boost British food production. Known as the MOM, the tractors remained the property of the Ministry of Munitions but were leased to farmers who were able to buy them after the end of the war. The first Model F tractors also had a Hercules side valve engine but later models had a Ford power unit with trembler coil ignition and a water wash air cleaner which developed

23 hp when running on petrol and 20 hp on paraffin. Trembler coil ignition, also used for early Ford cars, was not very reliable in cold weather. An electric current was generated by a system of magnets on the flywheel which rotated close to a series of fixed primary coils. The generated current was then passed into the trembler coils, increasing the voltage and delivering it to the sparking plugs. The moving parts of the engine splashed oil around the cylinder block and small scoops on the bearing caps lubricated the big end and main bearings. Failure to maintain the correct oil level in the sump was asking for trouble and standing the tractor on a slope with the engine running could starve the front or back big end and main white metal bearings of oil. The Model F's multi-plate clutch, which ran in oil, also served as a brake and the three forward and one reverse gearbox transmitted power through worm and wheel reduction gears to the steel wheels. An industrial version of the Model F with solid rubber tyres appeared in 1923.

The Fordson Model N, which replaced the Model F in 1929, was made in Cork until production was transferred to the new Fordson factory in Dagenham in 1933. The

Model N, still with a water wash air cleaner, was painted blue. The colour was changed to orange in 1937 and from 1939 the tractors were green, perhaps in an attempt to provide some camouflage during the war years.

There were four versions of the Model N. The standard tractor had steel wheels and the later land utility model had pneumatic tyres. A tricycle-wheeled rowcrop model and the industrial version completed the range. The Model N, better known as the Fordson Standard, cost £156 in 1931 but the price was reduced by £21 in 1936 to promote sales during the years of farm depression. The reduced price Model N on cleated steel rear wheels cost less than a team of three horses and when ploughing at a rate of half an acre in an hour the tractor was said to do the work of eight horses. The Fordson E27N (the E referring to England) replaced the Model N in 1945.

6. The rear wheels could be moved in or out on the tricycle-wheeled Fordson Model N's rear axle.

The Caterpillar Tractor Company of Illinois was formed when the Holt Tractor Co, which had already registered the Caterpillar name, joined forces with the Best Tractor Co in 1925. Benjamin Holt had introduced the first practical steam-powered tracklayer in 1904 followed by a petrol-engined model in 1906 while Daniel Best was making oil-engined tracklayers in 1910. The Caterpillar Best Sixty with a four-cylinder engine and two-speed transmission, which developed 59 hp at the drawbar, and the 25 belt hp Caterpillar 2 Ton launched as the Holt T-35 in 1921, were in production when Holt and Best merged their companies.

Jerome Increase Case founded the JI Case Threshing Machine Co at Racine in Wisconsin to manufacture steam engines and threshing machines in 1842. By 1886 Case was the world's largest steam engine manufacturer. The first Case tractors were made in 1892 and an improved model appeared in 1895 but little more was heard of them until the introduction of the first transverse-engined Case Crossmount 10-20 in 1915. The tricycle-wheeled tractor had three different-sized wheels and normally only the left-hand rear wheel was used to propel the tractor. The first four-wheel Crossmount tractors appeared in 1916 and within three years the Crossmount 10-18, 15-27 and 20-40, with the model numbers denoting the horsepower available at the drawbar and belt pulley, were in production at Wisconsin in 1921. The 12-20 and 18-32 replaced the 10-18 and 15-27 in 1924 and the 25-45 superseded the 22-40 in the same year. These tractors became known as the Case Model A, K and T in 1927. The Case Model L and Model C with in-line engines replaced the Crossmount tractors in 1929.

The Associated Manufacturers Co (Amanco), formed in America in 1906, opened a British subsidiary in London in 1912 to import stationary engines. They became the UK distributors for Case tractors in 1924 and sold Case Crossmount tractors from their premises. The tractors, adorned with the American Civil War Bald Eagle badge first used on Case steam engines in the mid-1860s, were painted green and red until 1923, then the colour was changed to grey and flambeau red in 1939.

Having completed an engineering apprenticeship in 1836 the young John Deere opened a blacksmith's business in the small town of Grand Detour, Illinois. He made his first steel plough there in 1837 and had added

7. The Case Crossmount 12-20 had a transverse engine, a cone clutch, two forward gears and one in reverse. This particular tractor caused a problem for the owner while driving round the ring at an agricultural show.

cultivators and corn planters when he opened a new factory at Moline in Illinois in 1859. The famous leaping deer trademark appeared in 1876. The founder died in 1886 and the company showed little interest in tractors until they acquired the Waterloo Gasoline Engine Co of Iowa in 1918. Founded by John Froehlich, the Waterloo Gasoline Engine Co made their first tractor in 1896 and the paraffin-engined Waterloo Boy Model R with one forward and one reverse gear appeared in 1915. The improved Waterloo Boy Model N with a 12–25 hp twin-cylinder engine and two forward gears and one reverse was introduced in 1917. Harry Ferguson who became the Belfast agent for Waterloo Boy tractors in 1915 sold the red and green tractor as the Overtime Model N. Associated Manufacturers were the Overtime importers for the UK.

John Deere built the green and yellow Waterloo Boy Model N until 1923 when the 15–27 hp John Deere Model D with a horizontal twin-cylinder engine, a two-speed gearbox and roller chain final drive was introduced in 1923. North America was the main market for the Model D but a considerable number were sold in Europe between the two world wars with many still in use in the late 1940s. FA Standen in Ely and H Leverton in Spalding were selling the green and yellow tractors in the UK in 1935 when John Deere introduced the improved 24–37 hp Model D with a three-speed gearbox. An even more powerful Model D rated at 30–38 hp, which appeared in 1940, was made until 1953. Tractor horsepower figures at the time denoted drawbar and belt hp in that order. The John Deere Model C rowcrop tractor launched in 1927 was a serious competitor for the International Farmall F. The 19–25 hp Model C had front and rear power take-off shafts, a belt pulley and a pedal-operated mechanical power lift for the toolbar. The Model C, re-badged as the John

8. *The Moline Universal Motor Plow had an articulated steering system.*

Deere GP in 1928, was advertised as 'a two-plough tractor of standard design that plants and cultivates three rows at a time'. The first diesel tractors with the leaping deer badge appeared in 1949 but twin-cylinder water-cooled petrol and vaporising oil engines provided the power for most John Deere wheeled tractors until the late 1950s.

Tractor manufacturers were experimenting with four-wheel drive tractors and motor ploughs in the early 1900s. The 18 hp Moline Universal Motor Plow made at Illinois in 1914 was the first to have an articulated steering system. Allis-Chalmers, Crawley, Fowler and Ransomes all made similar self-propelled motor ploughs at the time. Fowler's Rein Drive motor plough was driven in the same way as a horse. The reins were used from a seat at the rear of the motor plough to engage drive, change gear and steer the machine. The Boon motor plough, exhibited by Ransomes at the 1920 Royal Show had a twin-cylinder 18–20 hp paraffin engine, a two forward and one reverse gearbox and a two-furrow plough. The Boon was later sold by Eagle Engineering in Warwick. Minimal cultivations were already possible in the early 1920s when the 'Once Over' tillage system appeared in America. Basically a two-furrow motor plough with a four-cylinder 25 hp paraffin engine, it had a vertical shredding rotor above each mouldboard leaving the soil ready for the drill.

The 65 brake hp and 40 drawbar hp 8 plow Twin City 40, introduced by the Minneapolis Steel & Machinery Co in Minnesota in 1912, was at the opposite end of the power scale. Popular with prairie farmers in America and Canada the Twin City 40 had a water-cooled petrol or paraffin engine and a single forward gear with a top speed of 2 mph.

9. The 8-plow Twin City 40-65, introduced in 1912, weighed 13½ tons. It had a four-cylinder petrol/paraffin engine and the petrol tank held 10 gallons.

Massey-Harris, who had previously sold tractors designed by Parrett and Wallis, introduced the 25 hp four-wheel drive General Purpose tractor in 1930. Made for six years, the Massey-Harris GP, as it was known, had two forward gears and one reverse and its pivoting rear axle improved traction on uneven ground. Crawler tractor development owed much to the military tank. Benjamin Holt and Daniel Best were among the leading tracklayer pioneers in America in 1910. Clayton and Shuttleworth, who held a similar position in Britain, introduced a crawler with clutch and brake steering in 1918 and within ten years the International Harvester Co was making the 10–20 TracTracTor crawler in America.

Diesel engines were gradually coming into use in early 1900s but it was another twenty years before they were considered suitable for farm tractors. The Lanz single-cylinder two-stroke semi-diesel Bulldog tractor was an early leader in this field. Meanwhile, a few paraffin-engined models, including the Clayton Crawler and Peterbro' wheeled tractors, were gradually replacing the horse in the UK. The 35 hp Clayton crawler introduced by Clayton and Shuttleworth in 1916 had a Dorman

paraffin engine, two forward gears and one reverse and a steering wheel instead of the more usual steering levers. Peter Brotherhood at Peterborough introduced the 30–35 hp Peterbro' tractor with a four-cylinder paraffin engine, a cone clutch and a two forward and one reverse gearbox in 1920 for which it was awarded a bronze medal at that year's Tractor Trials in Lincoln. The engine was unusual in that some of the exhaust gases were passed to the carburettor to heat up the ingoing air as it was drawn into the cylinders.

Thirty-two tractors from North America, Britain, France, Germany, Hungary, Ireland and Sweden took part in the World Agricultural Tractor Trials held in Oxfordshire in 1930. The North American entry included Case International Harvester and Massey-Harris wheeled tractors along with five crawlers from the 10–14 hp Caterpillar Ten to the 50–60 hp Caterpillar Sixty. Germany was represented by the 14–20 hp Mercedes-Benz diesel, a 15–30 hp Lanz Bulldog and the petrol-engined 35–45 hp Linke tracklayer. The French contingent included two Austin tractors with a 12½–20 hp petrol engine and an 11–15 hp model with a paraffin engine, together with the Citroën-Kegresse and four-

10. The Massey-Harris General Purpose, introduced in 1930, was one of the first practical designs for a four-wheel drive tractor.

wheel drive Latil haulage tractors with four-cylinder petrol engines. The Irish entry, a Fordson Model N, suffered from a cracked cylinder block and was withdrawn from the event. The semi-diesel 20–30 hp HSCS from Hungary and two Swedish Munktell tractors also took part in the Tractor Trials. British entries included wheeled and Roadless crawler versions of the Rushton, a single-cylinder Marshall and a two-cylinder diesel-engined McLaren entered by JH McLaren at Leeds. The Peterbro', Blackstone and Aveling Porter tractors entered by members of the Agricultural and General Engineers Ltd completed the list. The tractors were put through a series of drawbar, belt and field tests with ploughs and cultivators. A comprehensive report detailed the performance of each tractor

11. The Lanz Bulldog tractor had a hot bulb semi-diesel engine.

12. *The spade lugs on the Peterbro' steel wheels could be replaced with optional rubber pads for road haulage work.*

13. *The Saunderson Universal tractor with a twin-cylinder paraffin engine – another of the tractors taking part in the 1920 Lincolnshire tractor trials – ploughed three quarters of an acre in one hour.*

including fuel consumption, maximum belt and drawbar horsepower and the fuel cost per 100 horsepower hours of work. Fuel at the time cost 4p for a gallon of paraffin, 7p for petrol and 2½p for diesel oil.

Garretts of Leiston, famous for their steam engines, introduced a wheeled tractor in 1930 and a Garrett crawler with Roadless rubber-jointed tracks was added in 1931. Sales literature explains that the wheeled model, which was awarded a silver medal at the 1931 Royal Show, had a 38 hp Aveling & Porter four-cylinder diesel engine, three forward gears and one reverse and a 20 gallon fuel tank. Garretts of Leiston were members of the Agricultural & General

14. The first Austin tractors were made in 1919.

Engineers (AGE) group and about twenty tractors had been made when the AGE group failed in 1932.

Farmers still committed to their horses in the mid-1930s began to give serious thought to buying a tractor when an expert of the day suggested that 'other than in exceptional circumstances a tractor should not exceed 15 hp at the drawbar'. For farmers who may not have understood the significance of drawbar horsepower they were told that a 20 drawbar hp tractor should be able to plough about an acre per hour on stiff land when pulling four 12 in wide furrows 5 to 6 in deep at 3 mph. This was surely quite a feat, but in those days soil was compacted only by the horse's hooves and the farmer's boots.

The Austin tractor with a modified Austin 20 car engine was introduced by Sir Herbert Austin in 1919. It was made at Longbridge in Birmingham and to promote sales to French farmers he built a factory at Liancourt in France. It seems that the Austin was too expensive for British farmers when compared with the Fordson and UK production ended in 1924. Austin tractors with more powerful petrol/paraffin engines developing 25 drawbar hp were also made in France from 1921. They had a three-forward and one reverse gearbox with a top speed of 3½ mph. Austin tractors taking part in the 1930 World Tractor Trials included 15 and 20 belt horsepower models with four-cylinder engines, three forward and one reverse gearboxes with a top speed of 3 mph and a belt pulley. Austin tractors with 25, 35 and 55 hp diesel

engines, introduced at the 1933 Paris Show, were made in France until the late 1930s.

Arthur Clifford Howard built a self-propelled three-wheel rotary cultivator with a Morris twelve engine in Australia between 1926 and 1930 and the four-wheel Howard DH22 tractor was made there from 1928 until 1952. The Howard Auto Rotary Hoe 22 Model DH – to give the tractor its full name – had a 22 hp four-cylinder water-cooled overhead-valve paraffin engine, five forward gears and one reverse and a power take-off shaft.

International Harvester made rowcrop, ploughing, orchard, industrial and fairway tractor variants of some models in the 1930s. The same four-cylinder 15–17 hp petrol and paraffin engine and similar transmissions were, for example, used on the Farmall F-14 rowcrop version, the I-14 industrial model, the O-14 orchard tractor and the F-14 Fairway version for turf maintenance. Independent brakes made it easier to cultivate between rows of potatoes and root crops and Farmall tractors in the late 1920s had these brakes which were operated by a system of cables from the steering linkage. The cable applied one of the rear wheel brakes when the driver turned the steering wheel to help him make sharp turns at the headland.

Ransomes, Sims & Jefferies made their first tractor with a 20 hp petrol engine in 1903 and a cumbersome

35 hp model, which was about 18 ft long, 11 ft high and weighing 10 tons, appeared ten years later. The first MG garden cultivators were made in Ipswich in 1936 and Ransomes & Rapier, also in Ipswich, made crawler tractors for a while in the mid-1930s. The first one with a twin-cylinder 15 hp engine weighing a hefty 2 tons appeared in 1934, followed in the same year by the RT50. The RT50 had a four-cylinder 80 hp Dorman-Ricardo diesel engine, six forward gears and one reverse and Roadless rubber-jointed tracks that could plough ten acres in a day with a six-furrow plough.

15. A system of cables connected to the steering linkage operated the steering brakes on some mid-1930s McCormick Deering Farmall tractors.

Most tractor manufacturers turned to armament production during World War Two but a considerable number of Fordson Model Ns along with a few other makes were also made during the conflict. Roadless Traction's war effort included converting Fordson Model N tractors for the Air Ministry. They were supplied with rubber-jointed full tracks and an extended fore-carriage, which not only turned the Model N into a half-track, but also added stability to tractors with a front-mounted crane or winch. With Europe at war, the British needed more tractors to boost home food production and the American Lend-Lease Act in 1941 enabled the USA to lend munitions and tractors to Britain and its allies. Repayments with 2 per cent interest were deferred until the end of hostilities when the sum was to be repaid over a period of fifty years. The Lend-Lease tractors, which included the Minneapolis Moline GTA, the International Harvester W6, Case and Oliver 80, were allocated by the local War Agricultural Committees (War Ags).

There were approximately 30,000 tractors at work on British farms in the early 1930s and by 1937 about one third of the annual production of 18,000 farm tractors in Britain was exported. The UK tractor population had risen to about 57,500 tractors in 1939 when this total included 43,000 Fordsons, 10,000 International Harvester tractors and 1,000 Ferguson Model As, together with another 3,500 from Europe and America.

Several new British manufacturers appeared in the immediate postwar period and between them they produced 58,000 farm tractors and 27,600 market garden tractors in 1947. The influence of American tractors was still strong in Britain but David Brown, Ferguson, Fordson and Marshall and Nuffield soon gained ground in a marketplace where demand still exceeded supply. The 1948 Agricultural Machinery Census recorded a total of 261,180 tractors including 14,800 tracklayers on British farms and 517,000 working horses compared with 724,000 horses in 1939.

The running costs of farm tractors have always been of interest to those who pay the bills. Statistics for a ten-year period from 1919 found that the average cost of running a tractor was £85.5s.5d per year or 2s.7d (12½p) per hour, and operating costs, with the tractor written off over ten years, were 3s 3½d (16½p) per hour. Other hourly costs included depreciation at 7¾d (3p), repairs and overhauls 5¼d (2½p), fuel 1s.6d (7½p) and labour was charged at 8½d (3½p).

Chapter 2

Allis-Chalmers – Dutra

Allis-Chalmers

The Allis-Chalmers Manufacturing Company was formed in 1901 when the American partnership of Fraser and Chalmers went into business with steam engine manufacturer Edwin Allen. The steel-wheeled 10-18 made at Milwaukee in Wisconsin in 1914 was the first Allis-Chalmers farm tractor. The model number indicated that 10-18's opposed twin-cylinder paraffin engine developed 10 hp at the drawbar and 18 hp at the belt pulley. When the United Tractor & Equipment Co of Chicago, which made the United tractor with a Continental side-valve engine in the 1920s, went out of business in 1932 the tractor became the Allis-Chalmers Model U. Allis-Chalmers later used their own overhead-valve four-cylinder paraffin engine for the 28–34 hp Model U with a four forward speed and one reverse gearbox and a hand-lever operated hand brake. It had the distinction of being the first tractor to have pneumatic tyres and from the mid-1940s electric starting and lights were optional for the Model U which was made until 1950. Variants included the tricycle front-wheeled Allis-Chalmers Allcrop, later known as the Model UC rowcrop tractor. The 'Hot-Rod' Model U was used in America as a publicity stunt for tractor racing.

Allis-Chalmers opened a distribution depot at Totton near Southampton in 1936 and the 51 hp four-plow Model A was introduced in the same year. Weighing 4½ tons the Model A had a 28 gallon fuel tank and a four forward and one reverse gearbox with a top speed of just under 10 mph. The mid-1940s range of Allis-Chalmers tractors included the A, B, C, U, WF models

16. The Allis-Chalmers Model U was one of the first farm tractors to have pneumatic tyres.

17. Introduced in 1940 and made for the next ten years, the 16 hp Allis-Chalmers Model C was advertised as a two-row rowcrop tractor.

and the tricycle-wheeled WC rowcrop model. The Model B was launched in 1938 and the Model C – introduced to American farmers in 1940 with a 16 hp engine and hydraulic lift – had mid- and rear-mounted toolbars and adjustable wheel track settings. The first Model B tractors with a Waukesha side-valve petrol/vaporising oil engine had a high fixed front axle but this was soon replaced with the livelier 15.5 hp Allis-Chalmers BE overhead-valve engine and adjustable wheel track front axle. The tractor soon became popular in Great Britain and considerable numbers were imported from 1938, and later under the World War Two Lend-Lease programme. When Lend-Lease ended in 1942 British farmers had to wait until 1947 before they could to buy a new Allis-Chalmers Model B tractor. Between 1947 and 1949 Model B skid units with a more powerful 22 hp engine were shipped from America to Totton where they were fitted with electrical equipment, wheels, and tyres. The Model U wheeled tractor and the new 37.5 drawbar hp HD 5 crawler were also available in the UK.

A few Allis-Chalmers Model G tool carriers, made in America between 1948 and 1955, were sold in the UK. It was a specialist rowcrop tractor with a rear-mounted

8 hp four-cylinder water-cooled Continental petrol engine, a four-speed gearbox with a low first gear, adjustable wheel track and a mid-mounted tool frame. The tractor was advertised as 'having enough power for a 12 inch plough and to pull trailed equipment from the rear drawbar'.

The Allis-Chalmers Manufacturing Co at Totton bought the obsolete Sale-Tilney factory at Essendine in Lincolnshire in 1950 in order to make engines for the Model B which was assembled at Totton. Output had peaked at about 1,000 tractors per year when production moved to Essendine where Allis-Chalmers also made All-Crop 60 combine harvesters and Roto-Balers. The English Model B had a 22 hp overhead-valve vaporising engine, a three forward and one reverse gearbox with a top speed of 8 mph, independent brakes and adjustable wheel track. With its high ground clearance, narrow waist and manually operated mid- and rear-mounted fully floating toolbars the Model B was the ideal tractor for rowcrop work. Advertised as a 'tractor with a two-minute toolbar' it was claimed that the mid-mounted toolbar could be wheeled into position and attached or removed very quickly.

Lawrence Edwards & Co at Kidderminster introduced the Hingley hydraulic three-point linkage conversion kit – priced at £25 – for the Model B toolbar in 1949. This may well have prompted Allis-Chalmers to offer their own optional three-point linkage in 1951 when the basic tractor cost £335. Other extras included electric starting, an 8 in diameter belt pulley and power take-off. The last Model B tractors with an optional Perkins P3 diesel engine and four forward speed gearbox were made in 1954. The same four-speed transmission and choice of a petrol, vaporising oil or Perkins P3 engine were used for the Allis-Chalmers D270 when it superseded the Model B in 1955. A live power take-off

18. The American-built Allis-Chalmers Model G, made between 1948 and 1955, was powerful enough to pull a 12 in mouldboard plough at 3 mph.

19. The Allis-Chalmers Model B with its Persian orange paintwork has a top speed of 8 mph.

20. There was a choice of a four-cylinder petrol or vaporising oil engine or a three-cylinder Perkins P3 diesel engine for the Allis-Chalmers D272. Electric starting for the spark ignition engines, hydraulic linkage, power take-off, belt pulley and powershift rear wheels were included in a long list of extra equipment.

21. The Allis-Chalmers ED40, with eight forward and two reverse gears, had a top speed of 13 mph.

22. The model numbers of the Allis-Chalmers HD7, HD10 and HD14 crawlers were advertised as the number of furrows the tractor could pull but this number of furrows probably did not apply on the heavier soils in the UK.

was the main improvement on the restyled D270. A 30 hp petrol engine, 26 hp vaporising oil engine or a 31 hp Perkins P3 were the options for the Allis-Chalmers D272 introduced in 1957. Electric starting, belt pulley, power take-off and hydraulic linkage were not included in the basic price of the petrol and vaporising oil tractors but an electric starter was included in the basic price of £514 for the diesel tractor. Specialist versions of the D272 included a narrow 3 ft 8 in high model for orchard work and a high clearance rowcrop tractor with narrow tyres.

Sales literature for the ED40, launched at the 1960 Royal Smithfield Show, described it as 'Today's finest all-duty tractor... Handsome!... Versatile!... and Economical!' The 37 hp ED40 (English Diesel 40) was the last British-built Allis-Chalmers tractor. It had an improved Standard 23C four-cylinder diesel engine that had previously been used for the Ferguson 35 tractor. The ED40 also had an eight forward and two reverse transmission with a top speed of 14.5 mph and a live category 1 and 2 hydraulic system with an engine-mounted hydraulic pump. The new Selective Weight Transfer hydraulic system was claimed to minimise wheelslip even in the worst possible conditions. Engine power was increased to 41 hp in 1963 when the improved Depthomatic hydraulics with top link

sensing replaced the earlier weight transfer system. The last ED40 was made at Essendine in 1969 but the last tractors were not sold until 1971. Allis-Chalmers had acquired Jones Bales at Mold in 1961 and when Bamfords of Uttoxeter bought Allis-Chalmers in 1971 a range of pick-up balers and hay-making machinery was marketed with either Bamford or Allis-Chalmers paintwork.

Allis-Chalmers wheeled tractor production continued in America. The 100 hp barrier was breached in 1963 with the launch of the 107 hp two-wheel drive D21 and a turbocharged version that was added in 1965 put the D21 past the 130 hp mark. Ten years later Allis-Chalmers were still making a range of 40 hp-plus tractors in America but the 208 hp four-wheel drive Allis Chalmers 440 with a Cummins V8 power unit, launched in 1972, was made by the Steiger Tractor Company in North Dakota.

Allis-Chalmers' interest in crawler tractors began in 1928 with the purchase of the Monarch Tractor Company at Illinois. The 31 hp Model M, made between 1933 and 1942, was the first Allis-Chalmers crawler. Based on the Model U wheeled tractor, the four-plough Model M tractor had the same the 32 hp Allis-Chalmers paraffin engine, four-speed transmission and a steering wheel instead of the more conventional levers to control the steering clutches. Derivatives of the Model M included the wide track WM crawler and an orchard model with shields over tracks.

The HD series of crawler tractors introduced in 1940 were made for ten years. They included the HD7, HD10 and HD14 with three-, four- and six-cylinder supercharged GM Detroit two-stroke engines. The model numbers indicated the number of furrows the tractors could pull on American farmland but this was not necessarily the case on heavy land farms in the UK. A number of HD crawlers were imported by the British War Agricultural Committee under the Lend-Lease programme. The 60 drawbar hp HD7 and 86 drawbar hp HD10 were ploughing tractors while the 127 drawbar hp HD14 was mainly used by the construction industry.

Advertisements for the HD5, introduced in 1948, explained that the new tractor with a 37 hp two-stroke engine, five forward gears and one reverse had fewer adjustments and greasing points than earlier models and attention with a grease gun was only required every 1,000 hours. The HD9 and HD15, which were more suited to civil engineering and construction work with their two-stroke engines and respective maximum drawbar hp rates of 72 and 109, replaced the HD7 and HD10 in 1951. New Allis-Chalmers crawlers with four-stroke engines and clutch and brake steering available

in the UK in the late 1950s included the 66 hp HD6, the 99 hp HD11 and the 148 hp HD16. The HD6 cost £3,900 in 1959.

The Allis-Chalmers and Fiat construction divisions merged as Fiat-Allis at Essendine in 1974, KHD bought the Allis-Chalmers farm machinery division in 1985 and Fiat-Allis became part of the new Allis-Gleaner Corporation (AGCO) in 1990. New AGCO Allis two- and four-wheel drive tractors launched in America in the early 1990s ranged from the 45–65 hp 5600 series to the 175 and 195 power take-off hp 9600 series with a powershift transmission and advanced computer technology in the cab.

Aveling-Marshall

The British Leyland Special Products Group adopted the Aveling-Marshall name when they bought the Marshall-Fowler business in 1968 and continued production of Marshall crawlers at the Britannia Works at Gainsborough in Lincolnshire. The Aveling side of the company started as the steam engine manufacturer Aveling & Porter in the mid-1800s while Aveling-Barford, established in 1933 to make road rollers, was acquired by British Leyland in 1968.

The buttercup yellow Track Marshall 56, 75C and 90 crawlers were all of a similar weight and size with 56 in track centres. The model number indicated the horsepower. The Track Marshall 56 had six forward

23. The Aveling-Marshall Track Marshall 90 with a Dowdeswell plough in the mid-1970s.

gears and two reverse while the TM75C and TM90 had five forward and three reverse gears. Optional equipment included hydraulic linkage, power take-off and a weather cab. The 125 hp six-cylinder Fowler Challenger 33 with six forward and four reverse gears, later renamed the AM140, completed the 1975 Aveling-Marshall crawler range.

Buttercup yellow paint and the Aveling horse symbol were also used for the AM100 and AM105, which were launched in 1975 with a six-cylinder 100 hp Perkins and 105 hp Ford engine respectively. The AM120, also with a six-cylinder Ford engine, had a five forward and one reverse gearbox; a single-lever hydraulic steering system was added in 1976. The new tractors had a dual clutch, power take-off, three-point linkage and a quiet cab with optional air conditioning.

Charles Nickerson bought Aveling-Marshall from British Leyland in 1979 (page 171), revived the Track Marshall name and production of the Track Marshall 135 and Britannia crawler tractors continued at the Britannia Works.

Bean

Mr Bean made a self-propelled toolbar with a tiller-steered front wheel for use on his smallholding and a small number of Bean toolbars were made at an aircraft factory in Blackburn in 1945. Humberside

Agricultural Products at Brough in Yorkshire took up production later that year and the tricycle-wheeled toolbar was made there for about a decade. The Bean toolbar, which cost £260, had an 8 hp side-valve Ford industrial engine with an electric starter, a three forward and one reverse gearbox, a Ford 10 cwt van rear axle and independent brakes.

The four-wheel tiller-steered version of the Bean introduced in 1950 cost £295, a four-row hoe unit was £45 and a five-row drill unit was £68. Other attachments included cultivator tines, a fertiliser spreader and a crop sprayer. A sales leaflet explained that the Bean toolbar would 'drill, hoe, top dress or spray six 12-inch rows of crop with a greater degree of accuracy and uniformity than has ever before been attained'. Furthermore, the Bean rowcrop tractor was so easy to handle 'that the driver will not be tired by teatime and can carry on working without fatigue'.

Thomas Green at Leeds manufactured the Bean toolbar from the mid-1950s until 1959 when Strathallan Engineering & Construction Co at Auchterarder in Perthshire introduced a new version of the tricycle-wheeled Bean toolbar. Known as the Bean self-propelled power unit the tiller-steered machine had a 16 hp twin-cylinder Petter diesel engine with an electric starter, a three forward and one reverse gearbox, an oil-immersed rear axle and inboard

24. *The Bean self-propelled toolbar was used for drilling, hoeing, cultivating, spraying and spreading fertiliser.*

25. The early 1960s Strathallan Engineering version of the Bean toolbar had a 16 hp Petter engine.

hydraulic brakes. Various attachments including hoes, seeder units and a sprayer were made for the hydraulically raised and lowered mid-mounted toolbar. The Strathallan Bean was still made in 1979 when it was shown with a six-row band sprayer at that year's spring sugar beet demonstration.

The Bean Equipment Company at Brays Lane in Ely exhibited the self-propelled Bean Beaver tool carrier at the 1983 Royal Smithfield Show. There was an option of a 14 or 26 hp diesel engine for the Beaver, which had four forward gears, power steering and a power take-off. Designed for rowcrop work and bed planting systems, the driver, who was seated alongside the rear-mounted engine, had a full view of the tool carrier's hydraulically raised and lowered mid- and rear-mounted toolbars controlled by double-acting spool valves. The improved Mk III Beaver with a 24 hp Lister twin-cylinder diesel engine and a six forward and two reverse gearbox was expected to cost £6,500 when it was announced in 1984.

Belarus

Belarus crawler tractor production in the Soviet Union dates back to the 1920s and the first Belarus wheeled tractors were made at Minsk in 1946. The STZ 15/30 crawler was built at a factory at Stalingrad in the early 1920s and within ten years the identical KHTZ 15/30B

was being made at Kharkov in the Ukraine. An improved STZ3 crawler, also made at Stalingrad, appeared in 1937 but twelve years later it was redesignated the Kharkov DT54. From 1953 some of the tractors made at various factories in the Soviet Union were exported under the Belarus name – mainly to Eastern Bloc countries – by the state-owned 'Tractorexport' organisation. The ageing DT54 was replaced in 1963 by the Belarus DT75 crawler with a 75 hp four-cylinder diesel engine started by a petrol donkey engine with an electric starter motor.

The first Belarus DT75 crawler with a £3,600 price tag appeared in the UK in 1973 at the Power in Action demonstration in Suffolk. The crawler's qualities were demonstrated by Ken Tuckwell Ltd. An improved DT75M appeared in 1975. Rated at 90 hp it had seven forward gears and one reverse with an underdrive gearbox providing a speed reduction in the four lower gear ratios. Other features included clutch and brake steering, hydraulic linkage and a weather cab. The Belarus tractor range was widened in 1978 when the D75M became the Belarus D750 and a new wide track model was known as the D750B. Standard equipment for both tractors included an exhaust spark arrester, a spring suspension seat, a starting handle and a tool bag with a grease gun and an inspection lamp. A touch of luxury was added in 1980 when a Lucas radio was

26. The Belarus DT75 crawler was introduced to British farmers in 1973.

27. There was a choice of rubber, steel-reinforced rubber or steel tracks for the late 1990s Belarus VT 100 crawler.

offered as an optional extra. Crawler production continued at the Volgograd (previously Stalingrad) factory but demand for crawler tractors in the UK fell and the UMO Group imported its last Belarus DT750 crawler in 1981.

Belarus crawlers reappeared in the UK in 1996 when UMO Belarus at Letchworth in Hertfordshire imported the Volgograd-built VT-100 crawler. It had a turbocharged four-cylinder engine rated at 120 hp for normal drawbar work but by adjusting the fuel injection pump the engine output could be increased to 150 hp for heavy fieldwork. The specification included a pneumatically operated clutch, ten forward and five reverse gears, air-assisted brakes and steering and an air-conditioned cab with a suspension seat. There was also a choice between road-legal rubber-padded steel tracks or steel-reinforced rubber tracks.

Postwar production of Russian wheeled tractors recommenced in 1946 when models designed for forestry work were produced at Minsk and the first MTZ-2 wheeled farm tractors with 37 hp diesel engines were made in 1953. The initials MTZ were short for the new name Minsk Tractorni Zavod (factory) and the

number 2 in the tractor model denoted two-wheel drive. The MTZ-5 with a wider wheelbase superseded the MTZ-2 in 1957; with an improved hydraulic system, the MTZ-5 became the MTZ-5K. The 48 hp MTZ-5LS with optional half-tracks was added to the range in 1959.

The first Belarus MTZ-50 tractors with a 75 SAE hp diesel engine and a nine forward and two reverse gearbox were made in Minsk in 1961. The four-wheel drive MTZ-52 was, apart from its front-wheel drive, identical to the MTZ-50. Variants included the MTZ-50X for cotton field cultivations and the narrow track T-54B for vineyard work.

The MTZ-50 Super and four-wheel drive MTZ-52 Super wheeled tractors with 70 SAE hp long stroke diesel engines, first made at Minsk in 1969, were introduced to British farmers in 1970 by Satra Motors at Byfleet in Surrey when they became Belarus tractor importers for the UK. Both models had nine forward and two reverse gears, power-assisted steering, disc brakes, independent and ground speed power take-off, independent front suspension and an 'extremely comprehensive tool kit'. The front-wheel drive on the MTZ-52 Super was automatically engaged when the

28. There were two- and four-wheel drive versions of the 52 hp Belarus MTZ-52 Super.

29. The drive shaft to the MTZ-52 Super's front wheels was designed long before the days of farm safety regulations.

rear wheels started to slip. Engine power for the MTZ-50 Super and MTZ-52 Super was increased to 75 hp in 1973 and after completing a very thorough pre-delivery inspection Satra Motors (Soviet-American Trade Company) sold the tractors for less than £1,500 plus delivery to the farm. An advertisement for Belarus tractors advised potential purchasers that 'the MTZ-52 Super does more ploughing miles per hour than any comparable 75 hp tractor'.

Independent front axle suspension was a feature of the Belarus T40 Super and the T40A Super introduced to British farmers at the 1972 Royal Smithfield Show. Both tractors, with 50 hp air-cooled direct injection engines, were made at the LTZ Lipetsk Tractor Zavod. Advertisements suggested that the British countryside would make a nice rest for them as they were more used to Russian winters and they would ease their way through the

toughest farm work. The claim had apparently been proved by a group of British farmers who had accepted a challenge 'to bash a T40 to destruction' but failed in the attempt.

30. Power-assisted steering, front axle suspension and a full set of lights were standard on the 50 hp Belarus T40A Super.

Tractors were becoming more powerful by the mid-1970s and Belarus followed the trend when they introduced the 90 SAE hp MTZ-80 and four-wheel drive MTZ-82 in 1974. The Belarus T25 29 hp twin-cylinder, air-cooled two-wheel drive bi-directional tractor from the Vladimir Tractor Zavod was launched at the 1974 Royal Smithfield Show. The T25 with an eight forward and six reverse speed gearbox, category II hydraulic linkage, ground speed power take-off and a Lambourn quiet cab on rubber mountings cost £2,295. The tractor was equally easy to drive in both directions for rowcrop work or mechanical handling after relocating the seat over the battery box, reversing the steering mechanism and rearranging the foot pedals.

UMO Belarus, a British subsidiary of the Russian United Machinery Organisation, replaced Satra Belarus in 1976 as the UK distributor for Belarus tractors and machinery. UMO also sold the Russian-built 300 hp Belaz K-710 heavy duty tractor in the UK in the mid-1970s. The giant tractor had a twelve-cylinder diesel engine, a sixteen-speed gearbox and hydraulic linkage and was claimed to be powerful enough to pull a twelve-furrow plough or a 60 ton load at the drawbar.

The Belarus 400 and four-wheel drive 420 launched in the mid-1970s had a 58 hp direct injection air-cooled engine and the gearbox provided a range of ten forward and two reverse gears with an additional reverse lever for six of the forward speeds. The specification included live and ground speed power take-off, live hydraulics, hydrostatic steering, a quiet cab and a toolkit with thirty different spare parts. The front-wheel drive on the 420 was automatically engaged when wheelslip exceeded four per cent.

Belarus tractors were re-numbered in 1978 to bring the UK numbers into line with those used in other countries. MTZs 50, 52, 80 and 82 became the Belarus 500, 520, 800 and 820 respectively and the T25 became the Belarus 250. Tractors sold in Britain the same year were supplied with a Duncan or Retford Sheet Metal Company quiet cab.

31. The Belarus 1500 with a 150 hp six-cylinder engine cost £19,978 in the late 1970s.

UMO Belarus were aware that tractor drivers had come to expect in-cab entertainment in the late 1970s when they advertised the new V-6 engined Belarus 1500. The banner headline read, 'If at £19,978 the 165 hp engine under the bonnet looks to give you all you need in a articulated tractor then for an extra £63 – the price of an optional Lucas radio – we can make the price sound positively sweet music.' The 1500, with three forward gears and one reverse in each of four hydraulically selected ranges, had power steering, category III hydraulics, a two-seat air-conditioned cab and a compressor which could be used to inflate the tractor's tyres. Sales competition in the UK was intense in the early 1980s when the Belarus tractor range included 29–165 SAE hp wheeled models and two versions of the 750 crawler. Like some other companies, UMO Belarus offered free gifts to help sell their tractors. A complimentary seven-day holiday in Russia was the carrot offered to farmers in 1981 when they bought a 50 hp-plus Belarus tractor.

Advertised as luxury models with greater driver comfort, repositioned controls and improved steering the first of the new Belarus D series tractors appeared at the 1982 Royal Smithfield Show. The 70 hp 560D/562D and the 90 hp 860D/862D had similar specifications with live hydraulics, hydrostatic steering, disc brakes, engine and ground speed power take-off and a luxury quiet cab. For farmers requiring a medium-powered tractor with a wide range of gears the 560D/562D, with a nine forward and two reverse gearbox, fitted the bill and the 860D/862D had eighteen forward and four reverse gears with a top speed of 21 mph.

Belarus launched the 100 hp MTZ-100 and made their two millionth tractor in 1984. The 110 hp Belarus 1060 and four-wheel drive 1062 added later in the year were very similar to the earlier 90 hp 860 and 862, apart from the more powerful turbocharged engine. The articulated Belarus 1507, introduced in 1986, was much the same as the earlier Belarus 1500 but a 24-volt electric starter motor replaced the very dated donkey petrol engine previously used to start the engine.

The Belarus 570 and 572 with 70 hp direct injection diesel engines were launched in 1988 and the 60 hp 540 and 542 with twelve forward and two reverse speeds were added in 1989 to fill a power gap in the Belarus tractor range. Tractors with the letter H after the model number indicated that it had the optional Scandinavian-designed Belarus-Hydrotronic hydraulic linkage control system used on some Russian-built tractors sold to farmers in the UK.

The introduction of the 570SB/572SB with eighteen

32. The standard specification for the Belarus 862 introduced in 1983 was as good as that of many of its more expensive competitors.

33. The 1025 was one of the first Belarus tractors made at Minsk with the new red and black colour scheme.

forward and four reverse gears was the main change in 1991 when the collapse of the Soviet Union resulted in Russian manufacturers exporting tractors under their own names. Those made at Lipetsk and Vladimir were sold in the UK with a UMO badge and a red, black and silver colour scheme while tractors made at the Minsk factory retained the Belarus badge and orange paintwork.

The 1993 Royal Smithfield Show was a busy one for UMO Belarus. The new UMO 250/252 and the first of the new Belarus 900 series tractors still with orange paintwork were introduced along with improved models of the 540/542 and the 115 hp 1082 in the new red, black and silver colour scheme. The Belarus 1082 had a twenty-four forward and eight reverse synchromesh gearbox, a draft, position and intermix hydraulic system and a flat floor cab. Features of the UMO 250/252 included an air-cooled twin-cylinder engine, eight forward gears, two reverse gears and hydrostatic steering. The 130 hp Belarus 1221, launched in 1994, had a six-cylinder turbocharged engine, a sixteen forward and eight reverse synchronised transmission and electro-hydraulic front-wheel drive engagement.

The Belarus 900/920, 950/952 and 1025 with 90, 205 and 115 SAE hp direct injection engines were launched in 1995 and the three millionth tractor was made at Minsk in the same year. Features of the 900 models included a synchromesh fourteen forward and four reverse gearbox, dry disc brakes, live draft, position and mix control category II hydraulics and a 540/1000 rpm power take-off with ground speed. The Belarus 1025 had hydrostatic steering, ground speed power take-off and quiet cab with optional air conditioning and there was a choice of a sixteen forward and eight reverse or a twenty-four speed synchromesh transmission.

The 1997 UMO Group catalogue included twenty-four models of Belarus tractor. They included a 14 hp articulated mini-tractor with a Briggs & Stratton engine, two- and four-wheel drive tractors in the 60–130 hp bracket, the VT100 crawler and two models of the articulated four-wheel drive Peter the Great. Introduced to British farmers in 1996 the Kirov Peter the Great had the option of a 184 or 257 hp engine and a 75-gallon diesel tank. It also had a sixteen forward and eight reverse powershift transmission, category III and IV hydraulic linkage and was capable of a turning circle with a radius of 25 ft.

Poor-quality finish on some tractors and high warranty claims caused the UMO Group to cease trading in 1999 when Belarus Tractors Ltd at Wokingham took over the UK distribution and marketed a range of two- and four-wheel drive models in the 35–135 hp bracket. They included the two- and four-wheel drive 900/920 and 950/952, the 70 hp 570/572 and the four-wheel drive Belarus 1221 with a six-cylinder 130 hp turbocharged engine, dual clutch, a sixteen forward and eight reverse gearbox and dry multiple disc brakes. A smaller range of six two- and four-wheel drive models, including the 920, 952, 1025 and 1221 was imported in 2001, but within a couple of years Belarus tractors were in short supply and importers struggled to survive due to the tractors' high prices.

Browns of Liversedge in West Yorkshire, a farming family who for some years had sold second-hand Belarus tractors, became the UK distributors in 2003 when they purchased the remaining stock of spare parts from Belarus Tractors Ltd. Supplies of new tractors were almost non-existent at the time but with improved build quality and new styling sales gradually increased and Browns advertised a new range of Belarus tractors in 2006. They included the 920.3, 952.3, 1025.3 and 1221.3 four-wheel drive

models with 90–140 hp Euro II compliance turbocharged engines under their red bonnets. The new range also came equipped with synchromesh or shuttle-shift transmissions, oil-immersed disc brakes, a heavy-duty front axle and an air compressor. Optional equipment included a synchro reverse shuttle gearbox, a mechanical creep speed gearbox and air conditioning. Advertisements at the time explained that the latest tractors had proven Belarus features with straightforward technology, no electronics to go wrong and were simple and economical to maintain.

BMB

Brockhouse Engineering – already well known for the pedestrian-controlled two-wheel BMB Plow-Mate, Cult-Mate and Hoe-Mate garden tractors – made the BMB, or Brockhouse President, between 1950 and 1956. The two-wheel garden tractors, originally made by Simplicity in America in the 1930s, were imported by British Motor Boats Ltd of London and later made at Crossens near Southport in Lancashire. Improved models of the 6 hp BMB Plow-Mate, 3 hp Cult-Mate and 1 hp Hoe-Mate were made between 1947 and the mid-1950s.

34. *Peter the Great, which weighed 16 tons, was made at Kirov near St Petersburg.*

35. The 10 hp Brockhouse President was made between 1950 and 1956. A narrow vineyard model was added in 1954.

The four-wheel Brockhouse President had a Morris 8 four-cylinder car engine with coil ignition and was rated at 16 hp when running on petrol and 14 hp on vaporising oil. Power was transmitted through a single-plate clutch, a three forward and one reverse gearbox, differential and spur gear final drive. A 6-volt electrical system – later upgraded to 12 volts – as well as independent brakes, an adjustable 40–72 in wheel track and a swinging drawbar were included in the basic price of £239.10s. Optional extras included a belt pulley, a 1⅛ in diameter power take-off shaft and hand-lift or hydraulic linkage toolbar. The hydraulic unit was bolted on to the side of the tractor and oil from the gearbox was pumped to a pair of external rams used either independently or together to operate the three-point linkage and the mid-mounted tool bar. A simple depth-limiting device with two metal blocks fixed to the mid-mounted toolbar and supported on the front axle controlled the working depth of hoe blades and cultivating tines. Screw handles were provided to adjust the working depth and the pivoting action of the front axle helped to maintain an even depth when working on uneven ground.

Sales literature explained that in exhaustive tests the Brockhouse President had given an astonishing performance out of all proportion to its size; the tractor's wasp waist and wire mesh driving platform afforded wonderful visibility. An advertisement for the President with the banner headline 'You haven't got money to burn' suggested that 90 per cent of the day-to-day work on any farm was light work that only required such a tractor. It added that many farmers were turning to the Brockhouse President, which was a light tractor with a high power-to-weight ratio, because it could do a lot of hard work at a very low running cost.

A narrow vineyard version was added in 1954 but with sales of this type of small tractor in decline the Brockhouse President was discontinued in 1956. The price was another deciding factor in its demise as farmers were able to get better value for their money by buying a second-hand Ferguson. A later version of the President (page 253) made by H J Stockton Ltd of London was exhibited at the 1957 Royal Smithfield Show.

Bray

Founded in the early 1900s, Bray Construction Equipment Ltd made loader shovels and dozer blades mainly for industrial use. Four-wheel drive loaders were added in the 1950s and the Bray Centaur tractor – designed by Essex farm contractor John Suckling and based on a Bray loader shovel skid unit – went into limited production in 1959. The four-wheel drive Centaur with live front and rear hydraulic linkages was really a high-output, self-propelled, bi-directional plough. The hydraulic pump was driven from the tractor's 96 hp six-cylinder Ford 590E diesel engine. The Centaur had a dry plate clutch, a six-speed gearbox with a mechanical forward and reverse shuttle and a top road speed of 14 mph. Left- and right-handed five-furrow ploughs, consisting mainly of Ransomes parts, were mounted on the front and rear hydraulic linkages and the Centaur ploughed up and down the field to the opposite hedge just like a steam balance plough in earlier days.

Four-wheel drive was becoming popular in the early 1960s and Bray Construction Equipment Co, now based at Feltham in Middlesex, returned to the tractor market in 1966 with the equal four-wheel drive Bray Four 10/60. Based on a Nuffield 10/60 the front wheels were driven by a central prop-shaft from a transfer box to a differential and epicyclic reduction unit on the front axle.

The Bray Four 4/65 with power-assisted steering, front diff-lock and self-energising disc brakes followed the launch of the new 65 hp Nuffield 4/65 in July 1967. The front-wheel drive layout was similar to that used for the Bray Four 10/60.

Although several major tractor manufacturers were making their own four-wheel drive tractors in the late 1960s the Leyland 384 was also given the Bray front-wheel drive treatment when it superseded the Nuffield 4/65 in 1969. The Bray Four 384 was the last four-wheel drive conversion made by the Bray Construction Co before they moved to Tetbury near Gloucester in 1971. The company was taken over by Matbro in 1973 and the new owners concentrated on the production of mechanical handling equipment.

Bristol

The first Bristol crawler was made at the Douglas motorcycle factory in Bristol in 1932 for Bristol Tractors at St James Street in London. It was less than three feet wide, weighed about one ton and used a gallon of petrol per hour. Like the more popular Ransomes MG crawler, it was used mainly by smallholders and market gardeners. An air-cooled 1,200 cc twin-cylinder horizontally opposed Douglas petrol engine with a twist grip throttle on the single tiller-style steering lever provided the power. The Bristol had a three forward and one reverse gearbox

36. The Bray Construction Co introduced the Bray Four 4/65 in 1967.

and power was transmitted through differential brakes to 7 in wide Roadless rubber-jointed tracks. An advertisement for the Bristol crawler explained that it could work day and night for less cost than the upkeep of two horses.

The Douglas Motor Company went into liquidation late in 1932 when production was transferred to the Bristol Tractors factory at Willesden. The first Bristol crawler tractors with an air-cooled British Anzani V-twin engine were made there in 1933. However, by the end of that year buyers had the choice of a twin-cylinder water-cooled horizontally opposed 7 hp Jowett petrol engine or a 10 hp Victor diesel engine. The Bristol, with overall track widths between 35 and 60 inches, had a three forward and one reverse gearbox, power take-off and a belt pulley. The Jowett Car Company bought Bristol Tractors in 1935 and production of Bristol crawler tractors with a Jowett engine and Roadless tracks moved to Idle near Bradford.

Bristol crawlers made between 1938 and 1942 had a Jowett petrol engine with either two or four horizontally opposed cylinders or a twin-cylinder Victor Cub diesel engine. The tractor had a three forward and one reverse gearbox and a flat radiator grille replaced the original bull-nosed bonnet. From 1942, Bristol crawlers had a 10 hp four-cylinder Austin petrol or vaporising oil engine still with the twist-grip throttle on the steering lever. In 1944, separate carburettors for petrol and vaporising oil were introduced in order to solve the problem of fuel contaminating the crankcase oil. This meant that the engine automatically switched to petrol when running at idling speed. The Austin car dealer HA Saunders bought the Jowett Car Company in 1945 and production of Bristol crawlers was moved to Earby near Colne in Lancashire. When the supply of Roadless Traction tracks came to an end in 1946 they were made in house and the last Austin-engined Bristol crawlers, with their design dating back to 1932, were made in 1947.

The new Bristol 20 introduced in 1948, had a modified low compression 22 hp overhead-valve Austin 16 car petrol engine with an optional Newage conversion kit for running on vaporising oil. Power was transmitted through a single-plate clutch, a three forward and one reverse gearbox and spur gear final drive. The tractor had multi-plate steering clutches controlled by two hand levers and independent foot

TAKES A UNIT-ATTACHED PLOUGH, TOOLBAR FRAME, ETC., AS WELL AS DRAUGHT IMPLEMENTS

NEW FEATURES
Hydraulic lift : TRU-TRAC plough with Furrow-width control ; finger-light clutch-brake steering with "hands-off" straight running; slow-running OHV Austin Industrial engine, developing 22 B H P at 1,500 R P M ; longer track units, plus scientifically calculated weight distribution, give smooth riding under all conditions at all speeds.

The New Bristol '20' has much more power than its predecessor, which had already established its reputation as the Tractor with the greatest drawbar pull for its size in the world. Yet it is still only 3' 3" wide overall, and only 14' from radiator to back of 2-Furrow Plough.

BRISTOL TRACTORS LTD., Earby, via Colne, Lancs

37. The rubber-jointed tracks on the 22 hp Bristol 20 were designed by Roadless Traction.

brakes were used for sharp turns. The standard Bristol 20 had a 29½ inch track centre and a special high-stability model for working on hillsides had 44 in track centres. The basic price of £480 for the Bristol 20 in 1948 included the power take-off but electric starting was an extra £22 10s.

The Bristol 20 was one of very few tractors made in Europe or America in the late 1940s with an optional bolt-hydraulic lift and three-point linkage. In this case the cost was an extra £47 10s. The hydraulic pump, constantly driven by the gearbox layshaft, supplied the hydraulic system with oil from the transmission housing. The power take-off shaft, which could be used with a belt pulley, was extended through the hydraulic unit on tractors with three-point linkage. Bristol Tractors sales literature for 'unit attached' (mounted) implements explained that when using a plough, cultivator or ridger the Bristol 20 could turn on its own axis and the front of the tractor could 'be driven right

38. Track guards for orchard and vineyard work were optional extras for the Bristol PD crawler.

up to the hedge before making a headland turn'. The leaflet added that the 'Bristol has always been the ideal tractor for cultivations – now perfect ploughing had been added to its other virtues.'

The modified 23 hp Austin A70 industrial power unit used during the latter part of the Bristol 20 production run also provided the power for its replacement, the Bristol 22, which appeared in 1952. An optional 23 hp Perkins P3 diesel was added in 1953. There was a choice of 7 or 10 in wide track plates and adjustment was provided for track width settings from 30 to 44 in. The optional bolt-on hydraulic linkage on the Bristol 22 could also be used with a wider range of equipment including single- and two-furrow ploughs, a toolbar, circular saw, post-hole digger and a winch.

The Bristol 22 became the Bristol 25 in 1956. Still with an Austin A70 engine it had conventional pin and bush tracks instead of the rubber-jointed tracks used on earlier models. The power take-off, belt pulley and hydraulic linkage were standard on the Bristol 25. The Bristol 25PC had a petrol engine, the 25VC ran on vaporising oil, the 25D had a Perkins P3 diesel engine and there was a special model for forestry work.

The more powerful PD series launched in 1959, with a 32 hp Perkins P3TA direct injection diesel engine and Ki-gass cold starting, was the last model of Bristol crawler. The PD (Power Diesel) tractors had a lever-operated single-plate clutch, a three forward and one reverse gearbox, clutch and brake steering and a Dunlopillo upholstered seat with a back rest. The was a choice of 7, 10 or 12 in wide track plates for the PD30, 33, 44 and 48 – the model number denoting track centre width in inches. The PD44 and PD48 were basically industrial models with a loader shovel or a dozer blade but track guards were available for orchard and vineyard work.

The Bristol Taurus made from 1964 to 1967 with a 46 hp direct injection Perkins engine and marketed by Bristol Saunders at Worcester could no longer be classed as a smallholder's tractor. The Taurus had six forward and four reverse gears, clutch and brake steering and hydraulic track adjustment. Optional equipment included a 540/1050 rpm power take-off, hydraulic linkage and a weather cab.

When the TW Ward Group bought Bristol Tractors

in 1970 the Taurus agricultural crawler was discontinued but the loader model remained in production for a while as the Track Marshall 1100.

Byron

Byron Farm Machinery Ltd at Walthamstow advised farmers and smallholders in 1946 that the new three-wheeled Byron Mk I tractor would be available in the following year. The advertisement explained that the tractor's single front wheel gave it excellent manoeuvrability and helped to keep the price at the competitive level of £260 on steel wheels and £280 on pneumatic tyres.

The Byron Mk I, like several other tractors of the day, had an industrial four-cylinder water-cooled Ford 10 petrol engine with 6-volt coil ignition and electric starter. The Ford 10 engine was a widely used power unit for small tractors as it was relatively inexpensive and spare parts were readily available. The Byron had a single-plate clutch, a three forward and one reverse gearbox with a top speed of 7 mph and independent

rear wheel brakes with a parking latch. The makers claimed that the Byron's mid-mounted toolbar, which could be raised and lowered with a hand lever from the driver's seat, made it ideal for rowcrop work and in average soil conditions the tractor could pull a two-furrow plough.

Sales literature offered owners handy tips on how to extend the working life of Mk I Byron tractors. They were advised to 'see that movable parts are greased sufficiently, check that the fixing nuts on the rear wheels are tight, never race the engine when it is cold' and to 'fill the radiator with very clean water'. Advice for the tractor's electrical system was to see 'that the battery is filled with distilled water, that the dynamo is charging and not to leave the ignition switch on when the tractor is not working'.

The Byron Mk II, which appeared in 1949, was the same basic machine with a few cosmetic changes, such as slimmed down mudguards to improve visibility for rowcrop work. The standard petrol-engined Mk II tractor with improved brakes and steering cost £247

39. The Byron rowcrop tractor had a 10 hp Ford industrial engine.

10s; some of the optional extras were a vaporising oil conversion kit, lights, belt pulley and steel wheels. Implements for the Byron tractor included mid- and rear-mounted toolbars, a plough and disc harrows. The Byron Mk II rowcrop tractor, complete with a 7 ft wide toolbar, cost £292 in 1949. This was considered expensive when compared with a petrol-engined Ferguson TEA20 for £325 or a rowcrop version of the E27N Fordson Major which cost £226. Consequently, there was little demand for the Byron and it went out of production in 1954.

Cabs

There was a time when on a cold and wet winter's day a tractor driver, crouched low down on the seat of his Fordson Model N, appreciated the protection provided by a coomb corn sack over his knees and another across his shoulders. In 'old money' a coomb sack was big enough to hold 2 cwt of barley or 2¼ cwt of wheat.

Some tractor drivers, especially those who had to spend long hours on a crawler tractor seat and were handy with a gas welding kit and basic carpentry tools, made a box-like cabin with glass or Perspex windows to protect themselves from the worst of the weather. Although these homemade cabs were usually draughty they were a huge improvement on a couple of corn sacks.

However, by the late 1940s several companies were marketing weather cabs for popular makes of tractor. A typical cab had a steel frame covered with sheet metal or canvas, removable doors and a safety glass windscreen. Weathershields sheet metal cabs for Ferguson, Ford and David Brown tractors were supplied with foot plates and fabric gaiters over the steering linkage. A sales leaflet reminded farmers that 'as it is essential to carry out farm work in all weathers the protection of a cab will reduce to a minimum the time lost through bad weather'.

Weathershields cabs for the Ferguson TE20 – with full front, side and rear vision that cost £28 ex-works – could be 'quickly and easily assembled on site'.

Sliding glass door panels were a feature of Victor Storm Guard cabs made by the Victoria Sheet Metal Company in Wellington, Shropshire. Prices started at £42, an optional electric windscreen wiper was an extra £4.9s and a roll-up rear curtain with a flexible plastic window cost £2. A two-door Victor cab for the MF35 had a detachable fibreglass roof and sliding side windows were provided so that hand signals could be made to other road traffic. The cheaper Victor Stormtop cab had a PVC cover over a tubular steel frame, two lift-off doors and a hand-operated screen wiper.

Sun-Trac cabs that came equipped with a hand-operated screen wiper cost £29.19s.6d; an opening front screen and a driving mirror were optional extras. An advertisement explained that Sun-Trac cabs provided 'complete protection against the roughest weather without restricting vision'.

The Minster cab, approved for the Farmall BM and BMD tractors, was made at Wimborne. The cab was bolted to the tractor mudguards and had a toughened safety glass screen, a canvas windshield above the fuel

40. Farmers could fit a Weathershields cab to a Ferguson TE20 without using a drill or a welder. An optional waterproof bonnet muff was claimed to keep the engine warm even in the coldest weather.

41.Some tractor drivers who were handy in the workshop made their own weather cab.

tank and low-level side windows. Scottish Aviation cabs were made in the late 1940s for Ferguson, Fordson Major, David Brown and Nuffield tractors and cost £29.15s delivered to the farm. Prices were similar for lightweight aluminium alloy cabs for David Brown,

Ferguson, Fordson, Massey-Harris, Nuffield and Turner Yeoman tractors.

Lambourn cabs with reinforced heavyweight rot-proof canvas covers and quick-detach doors were made in the early 1950s for Fordson, County and Roadless

42. Bristol Metal Components made the Sta-Dri cab on this John Deere 2120.

43. Lift-off doors and removable canvas cladding were features of Lambourn weather cabs.

tractors. These cabs were supplied in 'a knockdown condition' or, in modern-day terminology, as a flat pack. Lambourn also made an economy rear entry-only cab for the New Fordson Major which cost £22.10s, an optional side door adding £5.10s to the price. Sales leaflets explained that Lambourn cabs were 'quickly adaptable for all weather conditions in a matter of minutes – open sided in summer or fully enclosed for winter'.

Clyde-built Speedifit cabs – made by Innes Walker Engineering at Paisley in Scotland for Ferguson, Fordson, Nuffield, David Brown and International tractors – were also delivered in flat-pack form at a cost of £28.10s with carriage paid. Speedifit cabs were similar to an aeroplane cockpit canopy and could be attached to brackets on the mudguards in a few seconds. The driver could get on the tractor from the front or rear by tipping the cab and pulling it down again when seated on the tractor.

Portland cabs for the David Brown Cropmaster and Trackmaster were made in the early 1950s by the Portland Engineering Co at Halifax. Sales literature

noted that the Cropmaster with its wind-breaking scuttle was one of the warmest tractors to drive and the addition of an all-steel weather Portland cab provided real car comfort.

A range of metal cabs for most tractors with sliding windows, detachable doors and a roll-up rear curtain was made by Sta-Dri cabs at Bristol. Cabs for David Brown 850, 950 and 990 tractors had white fibreglass roofs and toughened window glass. The cheaper 'Tip-Top' canvas cab with a wrap-around windscreen had 'ample elbow room'. The Tip-Top cab was attached to the mudguards and folded back for 'exceptional ease of entry'. An advertisement explained that the cost of a cab and the ease of entry would mean 'no lost output or wasted hours for machinery and men'.

The Comfort Heater imported from America in the early 1960s by Farm Fitters Ltd at Gerrards Cross consisted of canvas sheeting secured around the tractor engine to direct warm air into the cab. Packed in a cardboard carton with full fitting instructions the

44. Farm safety regulations required all new tractors sold after September 1970 to have a safety cab or roll bar. Folding safety frames were permitted for tractors used in orchards, vineyards and low buildings.

Comfort Heater cost about £30 delivered to the farm. Ten years later Sirocco Tractor Equipment advertised the Easiheat with a similar engine shroud providing the driver with his own 'private heatwave in winter'.

Massey Ferguson introduced an optional steel-framed fibreglass cab with removable roof and doors for the 1965 range of Red Giant tractors. According to sales literature of the time, 'The new stylish-looking cab is free from drumming and rattles, completely rain proof and gives the driver maximum vision, comfort, operating space and convenience.'

The cab was an important sales feature in 1970 and manufacturers took steps to reduce noise levels. Massey Ferguson fitted a low-sound cab on the new MF595 and John Deere provided tractor drivers with a 'quieter operator protection unit' (cab). Lucas marked the introduction of improved and quieter cabs with the launch of a new in-cab stereo cassette player at the 1974 Royal Smithfield Show.

On average forty tractor drivers were killed every year in the late 1960s with overturning tractors cited as the cause of many of these deaths. This resulted in new legislation under the Health and Safety at Work Acts, which required all new tractors sold after 1 September 1970 to have a safety cab or rollbar tested and approved by the National Institute of Agricultural Engineering at Silsoe. Existing tractors driven by an employee were required to have a rollover protective structure (ROPS) in the shape of a safety cab or roll bar by 1 September 1977. The same regulations also required the provision of a step to aid entry to the driving platform. Sample cabs and rollbars were subjected to impact tests by swinging 2 ton blocks into the front, rear and sides of the cab. They also had to undergo a crushing test with the block lowered onto the cab roof. Only minimum deformation of the structure was allowed. The safety cabs gave protection from injury but the noise level

45.An air-conditioned safety cab is an integral part of the modern farm tractor.

inside the cab was increased and some manufacturers supplied ear defenders with new tractors to help deaden the noise from the engine and transmission.

Legislation finally solved the noise problem in June 1976 when the law demanded that all new tractors must be supplied with a quiet cab with a maximum noise level of 90 dBA. Isolating the cab from the chassis and using acoustic lining materials helped to reduce noise levels. Pendant pedals for hydraulic clutches and brake and power steering also contributed towards reducing cab noise to the required level. Some tractor drivers had a radio in the cab and sometimes farmers made their views known by commenting that although good money had been spent on a quiet cab there seemed little point if the driver had the radio at full volume.

46.Arm chair comfort, a computer screen and finger light controls make life easier for the driver in this John Deere cab.

Cabs have become more sophisticated over the years. By the turn of the century air-conditioned tractor cabs, often better appointed than a motor car, had comfort seats, push-button controls, a computer linked to the farm office and a satellite navigation system.

Carterson

Horace Carter and his son Philip designed and built the first Carterson Light tractor at Northwich in Cheshire in 1949. Front-, mid- and rear-mounted toolbars were made for the three-wheel tractor but stability problems resulted in the introduction of the four-wheel Carterson Light tractor. Except for its box section front axle it had the same specification as the three-wheeler.

The Carterson's kick-start side-valve Norton engine developed 8.5 bhp at 2,600 rpm and 16 bhp when running at its full throttle speed of 4,700 rpm. It had a three forward speed and one reverse gearbox with a top speed of 10 mph and an unusual clutch control mechanism returned the engine to tick-over speed when the pedal was depressed. The Carterson had a trailer drawbar and a spring-assisted hand lever was used to raise and lower the toolbar.

47. The Carterson was produced between 1948 and 1950. The tractor was made by Horace Carter and his son.

Steering was by a system of pulleys and wire rope, the wheel track was adjustable in 3 in steps from 30 to 36 in and independent rear wheel drum brakes gave the tractor a 4 ft 6 in turning circle.

The Carterson, which cost £245 ex-works in 1949, could also be used with a right-handed plough or a reversible plough with separate left- and right-hand bodies attached to the toolbar. A magazine article at the time explained that it had been 'the aim of the manufacturer to make it possible for growers to buy a tractor and its equipment for well under £300'.

Fifteen Carterson Light tractors were built at Northwich in 1949 and 1950. Another five sets of parts were made but they were sold unassembled and their fate is unknown.

Case

The JI Case Threshing Machine Co founded by Jerome Increase Case in 1842 made threshing machines and steam engines at a factory at Racine in Wisconsin. A petrol-engined tractor appeared in 1895 but little more was heard of Case tractors until the first of the range of transverse-engined 27–40 hp Crossmount tractors was introduced in 1915 (page 13).

Case had offices in London from 1919 when they introduced Crossmount tractors to the UK. From 1924 the agency passed to Associated Manufacturers (London) Ltd. The more conventional Case Model C and the Model L appeared in 1929 and the Model CC rowcrop tractor was added in 1930. The introduction of these tractors with enclosed roller chain drives to the rear wheels and grey paintwork marked the end of the era of Crossmount tractors.

The Model C with a 22–30 hp Case four-cylinder overhead-valve, water-cooled paraffin engine, three forward gears and one reverse, a hand clutch, roller chain drive to the rear axle, steel wheels and a belt pulley, was classed as a two–three bottom plow tractor. The Model CC was a rowcrop version of the Model C and from 1935 a foot pedal-operated power lift for the rear-mounted toolbar shaft-driven from the tractor engine was a factory-fitted option for the rowcrop tractor. Other Model C derivatives included the CO orchard model, the CV for vineyard work and a crawler version. The three–five bottom plow Case Model L, also with steel wheels, had a similar 22–28 hp four-cylinder Case paraffin engine, a hand lever-operated clutch, a three forward and one reverse gearbox, belt pulley and an optional power take-off. The Model C and Model L were restyled in 1939 with more rounded lines and a new Flambeau red colour scheme. The 40 hp LA, a more powerful four–five plow version of the Model L, had a hand clutch, four forward gears and one reverse with a top speed of 10 mph, power take-off and the option of steel- or pneumatic-tyred wheels.

The Case Model R made between 1936 and 1940 had an 18–20 hp four-cylinder Waukesha engine and a three forward and one reverse gearbox. Pneumatic tyres were an optional extra for the Model R and the tricycle-wheeled rowcrop Model RC.

The Model D with the new Flambeau red paintwork, launched in America in 1939, had a four-cylinder overhead-valve engine, four forward gears and one reverse and like earlier Case tractors it still had a roller chain final drive. Derivatives of the Model D included the tricycle-wheeled DC-3, the D-4 rowcrop model, the DO orchard tractor, the high-clearance DH and the DV for vineyard work. Associated Manufacturers were already importing the Case Model D, DC-4 and LA tractors when they moved to larger premises at the Palace of Industry in Wembley in

48. The 27 hp Case Model C with an enclosed roller chain drive to the rear wheels was introduced to American farmers in 1929.

49. The '4-5 plow' flambeau red Case LA tractor was made in America from 1939 to 1952.

1939. The Case DEX, an export version of the Model D and introduced in 1941, was specially modified with independent brakes, cast-iron rear wheel centres and a repositioned seat to meet the needs of British farmers who bought the tractor in considerable numbers.

Limited numbers of Case tractors with steel wheels were imported during the war years until the Lend-Lease arrangement ended. As the UK was short of dollars in 1945 the government announced its decision to halt the import of American tractors. Associated Manufacturers' stock of Case Model D, DC-4, and DEX tractors was exhausted by 1949 but they were still able to supply spare parts. Case tractor sales in the UK were transferred to Sale Tilney at Wokingham in Berkshire in 1954 when Associated Manufacturers went into receivership.

The two-plow Model S, introduced to American farmers in 1952, was one of the last tractors with the familiar Case Flambeau red paintwork. It had a four-cylinder 28–32 hp vaporising oil engine, a four forward and one reverse gearbox and belt pulley. Power take-off and three-point linkage hydraulics were optional extras. Model S derivatives included the SC rowcrop model and the SO orchard tractor.

The Case 500 with a six-cylinder diesel engine rated at 63 belt hp replaced the Model LA in 1953. The 500 had a hand clutch, a four forward and one reverse gearbox and roller chain drive to the rear wheels. The petrol, vaporising oil and LP gas versions of the Case 400 launched in 1955 had a dual range eight forward and two reverse speed gearbox, live power take-off and power steering. The four-cylinder 35 hp Case 300 with

50. The Case 2670 Traction-King with a twelve forward and four reverse gearbox was launched in 1974.

petrol or diesel engine and a twelve forward and three reverse gearbox for smaller farms was added in 1956.

The Case 1200 Traction-King launched in America in 1964 and the improved Comfort King, which replaced it in 1966, were not sold in the UK. Tenneco had acquired 91 per cent of the Case organisation in 1969 when the first of the 70 series Case Agri-King tractors appeared in America and within a year they were the outright owners of JI Case & Co. Tenneco added David Brown Tractors to their empire in 1972 and within a few years the Case name appeared on David Brown tractors sold in the UK.

Six models of the two-wheel drive Case Agri-King were made in America in the mid-1970s when, in common with other North American tractor makers, Case quoted power take-off horsepower in preference to engine power. The smallest Case 970 Agri-King was rated at 93 pto hp and the top-of-the-range 1570 launched in 1976 developed 180 hp at the power take-off shaft. The 970 Agri-King boasted a twelve forward and three reverse gearbox, hydrostatic steering, self-adjusting disc brakes, independent power take-off with a hand clutch and lower link sensing hydraulics. The Agri-King 1570 had a four-range gearbox with three power shift gears in each

range and its air-conditioned cab was isolated from vibration by rubber mounting blocks. A seat belt was standard equipment for the benefit of farmers who spent 'a lot of time in the cab on the Easy Rider 7-way adjustable swivel seat'.

The four-wheel drive Traction-King 2470 and 2670 launched in 1974 completed the mid-1970s Case tractor range. Rated at 176 and 221 pto hp respectively, they had a twelve forward and four reverse powershift transmission with an all-wheel steering system that gave the driver the choice of front-, rear-, four-wheel or crab steering. The turbocharged six-cylinder Case 2670 Traction-King, originally launched in America in 1974, and the 970 Agri-King with a safety cab were introduced to British farmers on the David Brown stand at the 1976 Royal Smithfield Show. Advertisements for the 243 hp Case 2670 with a £30,000 price tag explained that it had 'acre-eating power plus weather-beating traction' and the cab was 'the nearest thing yet seen to an airliner cockpit'. American farmers were able to buy even more horsepower in 1974 when Case launched the 300 SAE hp 2870 Agri-King.

The 90 series tractors superseded the 70 Series

51. The David Brown Case 2290 had an eight-way adjustable seat in its 'silent guardian' cab.

Traction-King in 1979. The five smaller David Brown Case 90 Series tractors with 48–103 hp engines and a twelve forward and four reverse transmission were made at Meltham. Seven more Case 90 series tractors with six-cylinder 120–273 hp engines first made in America in 1977 were introduced to British farmers in 1979 to meet a growing demand for more horsepower. The American 90 series included the 191 hp two-wheel drive 2590 and the four-wheel drive 4690 with six-cylinder turbocharged engines and semi-automatic twelve forward and four reverse powershift transmissions. According to sales literature the new American tractors had 'style, technical superiority and a hefty surge of raw power to make even the biggest operator feel at home in the cab'.

The David Brown name disappeared from the tractor scene in 1983. By 1985, after Tenneco had acquired International Harvester, the Case IH line up included sixteen models ranging from the 45 hp International 385L to the six-cylinder 145 hp turbocharged 1455XL. Case IH also introduced the improved 105 hp 1056XL in the same year. The specification included a sixteen forward and eight reverse synchromesh gearbox, independent power take-off and a new safety feature which automatically engaged four-wheel drive when the driver applied the brakes. Five Case 94 series tractors – from the 72 hp 1394 to the 291 hp six-cylinder 4894 with air-conditioned cabs – completed the 1985 Case IH tractor range.

Case Tractors Ltd at Meltham and Doncaster marketed thirty-four different tractors including the new Case IH 94 series tractors in 1987. The new 94 series tractors included the two-wheel drive 48 hp 1194, the 61, 72, 83 and 95 hp medium-powered tractors with two- and four-wheel drive and the flagship four-wheel drive 277 hp Case IH4894 with

52. The Case IH 1594 was one of ten models of the 94 series tractors launched in 1985.

equal-sized wheels and the latest electronic technology in the cab. The 94 series had a twelve forward speed gearbox and a Hydrashift semi-automatic change-on-the-move gearbox was optional for the medium-powered tractors.

The new Case IH Magnum 7100 series had been on sale in America for two years when it was introduced to British farmers in 1989. The four-wheel drive 7110, 7210 7310 and 7410 Magnums rated at 155, 180, 205 and 230 hp had an eighteen-speed full powershift transmission, a central drive shaft to the front axle and electronic linkage control (elc) hydraulics. The four-wheel drive 90 hp Case IH985 Turbo with a two-speed powershift gearbox, trailer brakes and XL cab was announced at the 1989 Royal Smithfield Show. It was advertised as a special edition for grassland farmers but within a year the ageing International Harvester 85 series had been replaced by the Case IH 95 series. There were seven 95 series tractors from the 44 hp 395 to the 995 with 90 hp under the bonnet. The two-wheel drive 395 had a two ratio powershift transmission and there was a choice of two- or four-wheel drive for the other 95 series tractors.

There were twenty-three Case IH models in the 17–264 hp bracket (including the long-lived 844XL) of the 1990 price list. They included seven 95 series tractors with 45–90 hp engines available in twenty-two different build combinations with a choice of three different cabs, two- and four-wheel drive and synchromesh or powershift transmission.

The specification of the 45 hp four-wheel drive Case IH844XL included a sixteen forward and eight reverse synchromesh transmission, self-adjusting inboard brakes and the International Sens-o-Draulic hydraulic system.

Three new 90–110 hp 5100 series Maxxums with sixteen forward and twelve reverse semi-powershift gearboxes capable of shuttle reversing and electro-hydraulically engaged front-wheel drive appeared in the same year. The lower link sensing Maxx-o-Draulic hydraulic system, which automatically matched hydraulic power output to the load requirement of the tractor, was a new feature of the Maxxum range.

A limited number of a no-frills Stockman's Special version of the 69 hp 695 and 82 hp 895, equipped with a basic cab and no heater, were made in 1992 when

Case IH announced that no discount was allowed on the tractor's £9,999 and £13,250 price tags.

A ready-made supply of five powerful four-wheel drive articulated tractors was the prize when Case IH acquired Steiger Tractor Inc of North Dakota in 1986. Marketed in America as the Case IH 9100 series with red paintwork and new model numbers they included the 190 hp 9100 (Steiger Puma 1000), 220 hp 9130 (Wildcat 1000), 280 hp 9150 (Cougar 1000), 335 hp 9170 (Panther 1000) and the 375 hp 9180 (Lion 1000). The 380 hp 16 ton articulated four-wheel drive giant Case IH 9280 made at the Steiger plant was the most powerful Case IH tractor yet seen when it appeared in 1992. It had a twelve-speed powershift transmission with a skip-shift mechanism which could be moved up two gears at a time to get the 9280 from 0 to 8 mph in about three seconds.

The six-cylinder Maxxum Plus 5150 with twenty-four forward and twenty reverse speeds launched in 1993 extended the power range to 125 hp. The smaller 90, 100 and 110 hp Maxxums had sixteen forward and twelve reverse synchromesh gearboxes and an optional creep speed box added another eight speeds in both directions.

Improved Maxxum Pro models and the new CS range were the first Case IH tractors with the Steyr pedigree following the company's link with the Austrian tractor maker. These tractors made their debut at the 1996 Royal Smithfield Show. Increased engine power, improved cab visibility and single lever control of both the range and powershift gearboxes were the main improvements on the Maxxum Pro series when they replaced the Maxxum Plus tractors. Features of the 150 hp Case IH CS150 included a 0 to 50 kph forward and reverse variable transmission and a computerised headland turn management system. The 94 hp CS94 had a sixteen forward and reverse powershift transmission and a four-speed power take-off.

The Steiger name reappeared on farm tractors when the first of six new articulated four-wheel drive 205–425 hp Case IH Steiger 9300 series tractors with turbocharged, after-cooled engines and a twelve or twenty-four speed synchromesh transmission was introduced in 1996. The addition of the 360 hp Quadtrac 9380 with four independently suspended rubber track units in 1997 widened the Case Steiger

53. The 310 hp articulated Case Steiger 9350 was among the range of 205–425 hp Case IH Steiger tractors launched in 1996.

range to ten big tractors with 'luxury operative protective structures' (cabs). Although more suited to wide open spaces in America some Case IH Steiger wheeled 'rowcrop' models and Quadtracs were sold in the UK in 1997. The more powerful American-built 17–261 hp six-cylinder Magnum Pro 7210 series replaced the Magnum 7200 tractors in 1997. The 170 hp 7210 was the only tractor in the five-model range without a wastegate turbocharger, which gave a 10 per cent increase in engine power for the 7220, 7230, 7240 and 7250 Magnum Pro series tractors. The full powershift twenty-four forward and six reverse transmission, power take-off, four-wheel drive and other settings were controlled from a multi-function driving console on the right of the Comfort Master seat in the Magnum Pro series Silent Guardian II cab.

The Case Maxxum or MX series was extended in 1998 to give nine models from the 80 hp MX80C compact to the Racine-built MX270 with 270 hp under the bonnet. The CX series with 50–100 hp Perkins engines and a manual or mechanical gearbox with a forward/reverse shuttle was added to the range of

Doncaster-built tractors in the same year. The letter M denoted that the tractor had a higher specification than the C range and S was included in the model number for Steyr-sourced models. The CS, CVX, MX Maxxum and MX Magnum series four-wheel drive tractors were current in 2001 when the Case IH catalogue ranged from the four-wheel drive three-cylinder 75 hp CS Compact to the six-cylinder MX Magnum 270 rated at 279 hp.

New Case IH Steiger articulated four-wheel drive tractors with 275 and 440 hp Cummins engines, including the 375 hp Quadtrac STX375 with larger rubber track units, appeared in 2000. They had sixteen-speed powershift transmissions with electronic shifting and optional Autoskip, which allowed the gearbox to miss alternate gears when moving off from stationary. Case IH added the 440 hp STX440 Quadtrac in 2001 and the 500 hp STX500 followed in 2003. New MXM models in the 120–200 hp bracket, with turbocharged and intercooled engines and a standard eighteen forward and six reverse semi-powershift transmission for the smaller models,

54. The Case IH Quadtrac made at the Steiger factory had a powershift transmission and four independently suspended rubber tracks.

55. The six-model 50–100 hp Case IH CX series, launched in 1998, was improved in 2000 with a stylish one-piece curved bonnet.

replaced the MX Maxxums in 2003. The 177 and 194 hp MXM models had full powershift and electronic engine management providing a power boost of up to 35 hp.

In 2005 Case IH tractors included the two- and four-wheel drive JX, JXC and JXU Maxxima models with 59–97 hp engines and synchro- or power-shuttle transmissions as well as the MXU and MXM Maxxum and the MX Magnum series tractors with engines in the 101–315 hp bracket. Five CVX models with 137–197 hp engines and constantly variable transmissions and six articulated Steiger STX tractors, including the STX450 and the STX 500 Quadtracs, completed the line-up.

Caterpillar

The Caterpillar Tractor Company of Illinois was formed when the Holt Tractor Company – which had already registered the Caterpillar name – and the Best Tractor Company merged in 1925. At this time, both companies were making petrol-engined crawler tractors.

The Caterpillar Twenty, made from 1927 to 1933, was one of the more successful early crawler tractors and some of these tractors – which had 25 belt hp four-cylinder petrol engines, gearboxes with three forward gears and one reverse, multi-disc steering clutches and band brakes – were sold in the UK. The petrol-engined Caterpillar Ten, Fifteen and Twenty tractors rated at 10, 15 and 20 drawbar hp took part in the 1930 World Tractor Trials held in Oxfordshire. The 25 hp Thirty and 59 hp Caterpillar Sixty also participated in the trials. A multiple-plate dry clutch, three forward gears and one reverse and clutch and brake steering were standard across the range. Prices in 1930 ranged from £290 for the Ten to £1,150 for the 8½ ton Caterpillar Sixty. The Caterpillar Twenty-Two was made between 1934 and 1939; the tractor had a 26 hp paraffin engine and proved to be more popular than the Twenty. Caterpillar decided that the future was in diesel power and introduced the four-cylinder four-stroke diesel-engined Caterpillar Diesel 60 in 1931 and like many later models including the Diesel 65 introduced in 1932 it was started with a 10 hp horizontally opposed

56. The Caterpillar D2, introduced in 1935, was the smallest of the Caterpillar D series tractors.

twin-cylinder donkey engine. Other 1930s Caterpillar diesel crawlers included the 41 hp 35 and the 82 hp Caterpillar 75 used mainly by the construction industry.

The first Caterpillar tractors arrived in Britain in 1915 and by the mid-1930s they were imported by Jack Olding & Co at Hatfield in Hertfordshire, H Leverton & Co at Spalding in Lincolnshire and LO Tractors of Couper Angus in Scotland. H Leverton & Co sold their first Caterpillar tractors in 1936 and were marketing agricultural versions – the D2, D4, D6 and D8 – in the mid-1940s. Nebraska tractor tests carried out between 1939 and 1941 on the D2, D4, D6, D7 and D8 recorded maximum power outputs of 30, 40, 78, 89 and 127 belt hp respectively. The D8 had a six-cylinder engine, as did the D6 from 1941 (3-cylinder before that), while the others had four cylinders. They were all started with a twin-cylinder donkey engine ranging from 10 hp for the D2 to a 24 hp engine on the D8.

The Caterpillar R2 and R4, only made for a few years in the early 1930s, were petrol-engined versions of the D2 and D4. The four-cylinder D2, with a 40 or 50 in track width, introduced in 1938 was made for the next twenty years. A few Caterpillar tractors including the more popular D2 were imported from the late 1940s. The December 1947 issue of the Farm Implement and Machinery Review reported that the D2 had a new engine developing 32 drawbar hp, an increase of 24 per cent compared with the previous power unit. The report added that with its extra power the D2 had an output of about 1½ acres in an hour when working at 3½ mph with a four-furrow 14 in plough.

The D4, D6, D7 and D8 tractors were launched at intervals from the mid-1930s and with improvements made from time to time some of them remained in production until the late 1970s. The D6B, for example, had five forward and four reverse gears and a 93 hp

57. The variable horsepower six-cylinder Caterpillar D6D was made between 1977 and 1986.

engine still started with a donkey engine. A special farm version of the 68 hp D4D designed for drawbar work appeared in 1964 with all of its five forward gears in the 2¼–5 mph speed range for cultivation work.

The American Caterpillar Company opened a spare parts depot near Leicester in 1950 and the first British-built Caterpillar tractors were made at Glasgow in 1958. The Caterpillar Tractor Company had depots in Glasgow, Leicester, Newcastle and London in the early 1960s and advertised the D4C as the operator's tractor. It had a 65 hp four-cylinder Caterpillar engine, a five forward and four reverse gearbox with a quick-change forward/reverse lever and lubricated for life track rollers. Most Caterpillar tractors were used by the construction and civil engineering industries in the mid-1970s but Levertons also sold agricultural versions of the Caterpillar D3, D4D, D5 (SA) and D6C with 62, 68, 90 and 125 hp engines and a Leverton hydraulic linkage.

The Caterpillar range of variable horsepower (VHP) crawlers launched in America in 1980 appeared at the 1982 Royal Smithfield Show. VHP models had an electric switching system linked to the transmission which increased engine horsepower when field conditions required more power at the drawbar. This was achieved with a solenoid in the injection pump, which increased the quantity of fuel supplied to the injectors when the driver changed up through the higher gears on the six-speed transmission. The horsepower ratings of the D5B and D6D (SA) were variable from 120–160 and 165–216 SAE hp respectively. The specification of the 97–125 Caterpillar D4E VHP exhibited at the 1983 Royal Smithfield Show included a 24-volt electric starter with an optional cold start system, an oil-cooled transmission clutch, hydraulic clutch and brake steering, a five-speed forward/reverse shuttle gearbox and 1,000 rpm power take-off. The D4E SA-T shared the D4E VHP specification with a standard 97 hp turbocharged four-cylinder direct injection engine. Other Caterpillar VHP crawlers included the D5B SA and D6D SA. The D5D developed 120 hp in gears 1 and 2 and 160 hp in gears 3, 4, 5 and 6. The D6D SA

58. A steering wheel was used to control the Caterpillar Challenger 65's hydraulic motors. Turning the wheel increases the speed of one rubber track while the other is slowed down.

59. The Challenger 35 with Mobil-trac rubber tracks was introduced to British farmers at the 1994 Royal Smithfield Show.

was rated at 165 hp in the two low gears and 216 hp in the higher gears.

Slow working speeds and problems with driving on the road were the main disadvantage of steel-tracked crawler tractors. The high speed Mobil-trac rubber-tracked Caterpillar Challenger 65 with a top speed of 18 mph introduced to British farmers in 1989 overcame the problem. The six-cylinder turbocharged 270 hp Caterpillar engine transmitted power to a ten forward and two reverse powershift gearbox and mechanical drive to a differential steering system. The rubber track belts reinforced with four internal flexible steel cables ran on pneumatic front idler wheels and steel driving wheels with four rubber-covered bottom rollers.

The improved 285 hp Challenger 65B and the new 325 hp Challenger 75 appeared in 1990 with both tractors having a ten forward and two reverse powershift transmission. The first Challenger 65 sold in the UK was to a 3,000-acre farm in Suffolk. The Challenger 65C and Challenger 75C superseded the earlier rubber track models in 1992. The main difference on the C series Challengers was the use of a

rubber-covered steel front idler wheel in place of the earlier inflatable rubber version which was liable to puncture on stony soils. The 70C and 85C, with 285 and 355 hp engines were added in 1993. An ADAS report suggested the cost of the Challenger was difficult to justify on farms under 1,200 acres and on even bigger areas it was desirable to operate shift work to make maximum use of the tractor.

The smaller Caterpillar Challenger 35 and 45 Mobil-trac rowcrop tractors were introduced at the 1994 Royal Smithfield Show. Rated at 205 and 235 gross engine hp with five track belt widths and five track centre settings, the new models with a large diameter driving wheel had a sixteen forward and nine reverse powershift transmission and differential steering. The 272 hp Challenger 55 rowcrop model was added in 1995.

When the 330 hp 75D replaced the 75C and the 272 hp Challenger 55 was launched in 1996 there were six Caterpillar Challengers, from the 212 hp 35 to the 330 hp 75D, on the British market.

Four improved E series Challengers – the 65E, 75E, 85E and 95E with 310–410 hp engines and modified transmissions – were introduced in 1997 and an agreement made with Claas that year resulted in the German company marketing Challenger tractors with Claas green livery in Europe. In a reciprocal arrangement Claas combines were sold in America with Caterpillar yellow paintwork. Four Challenger MT700 series tractors, introduced in 2001, with 255–320 hp C-9 Caterpillar engines had a sixteen forward and four reverse powershift transmission. Optional extras included a steerable three-point linkage and satellite navigation.

Within a few months the improved and restyled E series Claas Challengers were launched and Claas sales literature claimed their introduction to 'herald the arrival of user-friendly crawler tractors'. The electronically controlled Caterpillar engines on the rubber-tracked 65E and 75E were rated at 310 and 340 hp respectively. The 85E and 95E also developed 340 hp in first and second gear but this could automatically be increased to 375 and 410 hp when used in gears three to ten. The Challenger specification included a ten forward and two reverse powershift transmission, hydraulic differential steering and load sensing category III and IV hydraulic linkage. An eight-way adjustable driving seat and a buddy seat were provided in the air-conditioned cab. The 212–410 hp Challenger 35, 45, 55E, 75E, 85E and 95E rubber-track crawlers were in production in 2002 when Caterpillar withdrew from the agricultural market and AGCO bought the right to manufacture the Challenger rubber-tracked crawler.

The MT700 and MT800 series tractors were the first AGCO-badged Challenger crawlers sold in the UK. The four MT700 models with six-cylinder 235–306 hp Caterpillar engines made their debut at the 2002 Cereals event and the 500 hp MT 865 Challenger was launched at that year's Royal Smithfield Show. AGCO established a new world planting record in the Ukraine in 2003 when a Challenger MT 865 with a 13.35 m wide Horsch min-till drill planted 571.9 hectares in 24 hours.

Five new Challenger MT 800B models, launched in 2005 with 350–570 hp Tier III emission compliant six-cylinder Caterpillar engines had powershift transmissions with a top speed of 25 mph and electronic hydraulic linkage control. The 607 hp Challenger 875B, which cost £232,000 in 2006, was one of the most expensive crawlers on the British market.

Claas

The Claas Huckepack, a four-wheel self-propelled implement carrier made between 1957 and 1960, had a 12 hp air-cooled Hatz flat-twin diesel engine, a five forward and one reverse gearbox, diff-lock, power take-off, hydraulic linkage and adjustable wheel track. Attachments for the Huckepack included a 6 ft cut wrap-around combine harvester, a seed drill, a crop sprayer and a cutter bar mower. A separate 34 hp VW petrol engine was used to drive the combine cutting and threshing units which could be removed without difficulty to leave the tool carrier free for other work throughout the year. The Huckepack was also used with a load-carrying platform for haulage work and the driving platform placed above the left-hand rear wheel gave the driver a clear view when mowing grass or cultivating rowcrops.

Claas took no further commercial interest in tractors until 1990 when they co-operated with Schlüter in the design of a bi-directional Euro-Trac systems tractor (page 242), followed by public demonstrations of a prototype Claas Xerion multi-function tractor in 1993. With some resemblance to the Euro-Trac, the Xerion could be used with conventional front- or rear-mounted equipment or as a power unit for a root harvester, cultivator/drill combination, etc. With development work complete the first 200 hp Xerion 2000 tractors, which cost just under £100,000, were made in 1997. The range was soon extended with the addition of the Xerion 2500 and 3000 with six-cylinder turbocharged and intercooled 250 and 300 hp engines. The Xerion had a hydrostatic/mechanical transmission with eight mechanical ratios and stepless speed variation in each gear, front-, four-wheel and crab steering and front, centre and rear power take-off shafts. The cab, with electronic controls and a performance monitoring system, could be rotated hydraulically through 180 degrees to give the driver full vision and control in both directions of travel.

Although about eighty Xerions were in use worldwide by 1999 farmers were slow to accept the concept of a systems tractor. No new machines were sold between 2000 and 2003 but Claas engineers were busy designing an improved Xerion. The first production models, launched in 2004, had a more powerful 335 hp engine, a new constantly variable

60. The first 200 hp Claas Xerion tractors were made in 1997.

61. Claas marketed the 272 hp Caterpillar Challenger 55 with Claas green livery in Europe between 1997 and 2002.

62. After buying the Renault tractor factory at Le Mans, Claas exhibited a range of tractors with a light green colour scheme at the 2004 Royal Smithfield Show.

transmission (CVT), equal-sized wheels and a longer wheelbase. There were three alternative cab arrangements. The 3300 Trac Xerion had a mid-mounted fixed cab, the 3300S had a fixed forward-mounted cab and the cab on the 3300VC reverse drive tractor could be turned through 108 degrees.

Meanwhile an agreement with Caterpillar in 1997 resulted in Claas marketing the 212, 242 and 272 hp Challenger 35, 45 and 55 rubber-tracked crawlers in Claas green livery throughout Europe. Under the joint venture Caterpillar would market several models of the recently introduced Claas Lexion combines with Caterpillar yellow paintwork in America. This arrangement continued until 2002 when Caterpillar withdrew from the agricultural machinery market and sold their agricultural machinery interests to AGCO, thus ending the agreement with Claas to sell Challenger rubber-tracked crawlers in Europe.

Claas bought the Renault tractor factory in Le Mans in 2003 and French-built tractors including the

Atles, Ares, Pales, Fructus, Nectis and Celtis models, together with the Xerion 330, appeared in Claas livery at the 2004 Royal Smithfield Show. Optional transmissions for farm and for amenity work were available for the new four-model 68–101 hp Nectis range and the three smaller Pales tractors with 56–77 hp engines, at the lower end of the power range, had a twelve forward and reverse synchro-shuttle transmission. The new Celtis range with 70–100 hp John Deere engines launched in 2003 were the last Renault-designed Claas tractors. Most new Claas models could be supplied with the optional eDrive guidance system after it was introduced in 2004. Linked to GPS satellites eDrive took control of steering between headland turns while the tractor's headland management system took charge of the tractor and implement. With the exception of the 335 hp Xerion, the four Atles models – equipped with 232–282 hp Deutz engines and eighteen forward and reverse power shuttle transmissions current in 2006 – were the most powerful Claas tractors on the market.

63. The 1997 Clayton Buggi had a load sensing power take-off and a compressed air trailer braking system.

Clayton

The Clayton 4105 Buggi lightweight tractor and low ground pressure spray vehicle were launched at the 1992 Royal Agricultural Show by Lucassen Young of Stockton-on-Tees in Cleveland. The two- and four-wheel drive 105 hp Buggi with an all-round suspension had a John Deere turbocharged diesel engine and a ten forward and two reverse speed transmission. Other features included hydrostatic two- and four-wheel steering, a fully automatic trailer braking system and a moveable six-spline power take-off shaft driven by a hydraulic motor. Optional equipment included hydraulic linkage and hydrostatic transmission. A sprayer, fertiliser spreader or seed drill could be mounted on the rear load platform and the drawbar could be used for towing a trailer or for light cultivation work.

There were three versions of the Clayton Buggi on the market in 1994 with similar specifications to the original model including John Deere engines, re-locatable hydraulically driven power take-off, all-round coil spring and rubber shock absorber suspension and a 3.2, 3.8 or 4.2 m wheelbase. There was a choice of an 85 or 110 hp engine for the 4090 Buggi, the 4105 had a 110 hp or optional 120 hp turbocharged engine and the 4120 Buggi had a 120 hp power unit that was suitable for high-capacity de-mount sprayers and fertiliser spreaders.

An optional multi-function Pentronic steering unit was introduced in 1997 for the four-wheel drive Clayton 4105 and 4120 Clayton Buggi. When the tractor was used in four-wheel or crab steering mode a delay in the four-wheel Pentronic steering system allowed the front wheels to turn through ten or fifteen degrees before the rear wheels turned to match the angle of the front wheels. Other improvements included an air-conditioned cab, a load-sensing power take-off and a compressor for the air brakes and air controls on some liquid fertiliser sprayers. The four-wheel drive Clayton C-Trac 6700 and 6800 with 155 and 185 hp six-cylinder John Deere engines, a ten-speed synchromesh gearbox, four-wheel steering and the choice of three wheelbase lengths were added in 1997.

Multidrive Ltd at Thirsk acquired Lucassen Young in 1999 and for the next ten years or so they marketed a modified Clayton tractor badged as the Multidrive. The 2002 Multidrive range included two- and four-wheel drive models with 115, 125 and 155 hp engines and ten or twelve forward and two reverse speed gearboxes. In 2006, the four-cylinder 140 hp Multidrive 4140 and six-cylinder 185 hp Multidrive 6185 Pentronic tractors with powershift transmissions cost £57,975 and £64,000 respectively.

Cletrac

The Cleveland Motor Plow Company made the first Cleveland Model R crawler tractors at Cleveland, Ohio in 1916. The

64. Caterpillar and Cletrac crawlers including the Cletrac Model 15 were top-selling tractors in America in the early 1930s.

12–20 hp Model R had the distinction of being the first crawler tractor with controlled differential and brake steering instead of the more usual clutch and brake system. The Model R was steered by applying a lever to slow down one track while the differential automatically increased the speed of the opposite track to change the direction of travel. HG Burford of London marketed the Burford-Cleveland Model R in 1917 but it was quickly superseded that same year by the 12–20 Model H with one forward and one reverse gear. The Cletrac name appeared when the company changed its name to the Cleveland Tractor Company in 1918.

The Cletrac Model W, which replaced the Model H in 1919, was still being imported by HG Burford Ltd in 1924 when it cost £350. Other early Cletracs included the 16 hp lightweight Model F made between 1920 and 1922 while the K20 with a 24.5 hp overhead-valve engine, a power take-off and rear belt pulley was added in 1925. The four-cylinder 22–30 hp Cletrac 15 and six-cylinder 27–33 hp Cletrac 25 with Hercules petrol engines were made in the early 1930s. The Hercules Motors Corporation of Ohio was one of the world's largest manufacturers of two-, four- and six-cylinder petrol and diesel engines at the time. The orange Cletrac Model E introduced in 1935 with a choice of five

track centre widths from 31 to 76 in was also imported in considerable numbers by HG Burford Ltd. Other 1930s Cletrac crawlers included the BD diesel and 95 hp petrol-engined 80G with a six-cylinder Hercules engine. The BD, made between 1936 and 1956, had four forward and two reverse gears and when Oliver bought Cletrac in 1944 the tractors had yellow paintwork.

The Cletrac HG crawler with a four-cylinder 26 hp Hercules petrol engine, a three forward and one reverse gearbox and a top speed of 5¼ mph was the smallest of three early 1940s petrol- and diesel-engined Cletrac crawlers. The Cleveland Tractor Co recommended the 30–38 hp Model A series for medium-sized farms and the 38–50 hp Model B series with a petrol or diesel engine for 'large farms and for farmers who did custom (contract) work'. The Cletrac AG (gasoline) had a petrol engine and the diesel-engined Cletrac AD had an electric starter motor or optional pony (donkey) engine. After starting the pony engine a hand lever was used to engage a vee-belt drive to the starter pinion which turned the flywheel ring gear and hopefully the Hercules engine sprang into life. An advertisement warned American farmers that wartime regulations meant that it was only possible to buy a Cletrac crawler if they were able prove the need for a new tractor.

65. The 18–22 hp Cletrac HG crawler had a four-cylinder Hercules petrol engine.

The lightweight Cletrac General GG launched in 1939 was a tricycle-wheeled version of the Cletrac HG crawler. It had a 14–19 hp four-cylinder Hercules engine and a three forward and one reverse gearbox. The General GG, the first and only wheeled Cletrac tractor, was discontinued when the Oliver Corporation acquired the Cleveland Tractor Company in 1944. The new owners retained the Cletrac crawler range and the HG, with a 38 hp Hercules engine and a three forward and one reverse gearbox, was the smallest of three Oliver Cletrac models in the mid-1940s. When the White Motor Corporation acquired the Oliver Corporation in 1960 they dropped the General GG wheeled model but Oliver Cletrac crawlers were made for the next five years.

BF Avery & Sons at Louisville, Kentucky, who were making motor ploughs in 1915, acquired the production rights for the General GG wheeled tractor, which – with an improved Hercules engine – was badged as the Avery Model A. Blaw-Knox Ltd sold a few Avery Model A tractors to British farmers in the mid-1940s and the Avery Model A and Model V wheeled tractors, with a strong family resemblance to their Cletrac General ancestors, were still being made when Minneapolis-Moline bought BF Avery & Sons in 1951. The one-plow Avery Model V and the two-plow model BF with four-cylinder Hercules petrol engines were still current in 1954.

Cold Starting Aids

When atomised diesel fuel is sprayed into air heated to a temperature of approximately 550°C it burns rapidly and the expanding gases in the cylinder drive the piston downwards on the power stroke. Early diesel engines, known as hot bulb or semi-diesels, were started by using a blowlamp to heat an iron bulb extension on the cylinder head to raise the cylinder temperature high enough to ignite the fuel. Most diesel tractors made in the late 1940s and 1950s had an indirect injection diesel engine which was more fuel efficient than a direct injection engine but could be difficult to start in cold weather without the use of some form of starting aid.

A smouldering paper wick inserted in a holder and screwed into the cylinder head was used to help ignite the fuel when starting a Field Marshall tractor. In cold conditions a special cartridge was placed in the combustion chamber above the piston and the engine rarely failed to start when the cartridge was fired with a sharp blow from a blunt instrument.

An electric heater coil in the inlet manifold and a small hand-operated pump was a common form of cold start aid for multi-cylinder indirect injection engines. A small quantity of fuel from the main tank was pumped on to the red-hot coil and the resulting flame was drawn into the cylinders by the starter motor. The Fordson Dexta instruction book advised that this procedure could be repeated three times but if the engine still refused to start the driver should refer to the fault-finding chart! The Ki-gass cold start system used on the Ferguson TEF20 and some other diesel-engined tractors was used in the same way but a separate tank stored the fuel which was pumped on to the heater coil. The TEF20 also had a de-compression lever used to hold some of the valves open to reduce the load on the starter motor.

Many International Harvester tractors had a 'glow plug' pre-heater coil in each cylinder with a pilot unit on the instrument panel. The engine invariably refused to start when cold without pre-heating the cylinders and this procedure was often necessary when the engine was warm.

If all else failed the driver could always spray ether from an aerosol into the tractor's air cleaner. Most aerosols also contained an upper cylinder lubricant and the manufacturer of 'Easistart' claimed that it was amazing how many starts could be made with a single canister. The engine usually responded to this treatment but a cloud of black smoke and a loud knocking sound coming from the cylinder block could be a cause for concern. Some drivers were convinced their tractor had become addicted to it and the engine refused to start without its daily dose from an aerosol can! The truth was that worn cylinders and leaking valves meant that most of these engines were so short of compression that they would no longer start in any other way.

The transition to direct injection engines solved most cold starting problems and those with an in-line fuel injection pump normally had an excess fuel button for cold starting. A combined heater coil and automatic thermostatically operated excess fuel device in the inlet manifold was usually provided on engines with a distributor-type injection pump.

County

Ernest and Percy Tapp established County Commercial Cars at Fleet in Hampshire in 1929. Initially concerned with converting lorries to increase their load-carrying capacity they used their wartime experience gained in the development of lightweight tanks to design the first County CFT (County Full Track) agricultural crawler.

Introduced in 1948 the CFT crawler had an E27N Fordson Major skid unit with a petrol or vaporising oil engine. An optional Perkins P6 diesel engine was added in 1950. The Fordson Major's gearbox and a fixed spiral bevel on a countershaft transmitted power to the spur gear final drive which gave the CFT a top speed of 3.4 mph. The tractor was steered by a system of internal expanding brakes enclosed in a brake drum. Hydraulic linkage, power take-off, electric starting and lights were all optional extras. The County CFT, which cost £765 ex works in 1949, had five track rollers on each side and the standard 12 in wide track plates had a ground pressure of less than 4½ lb/sq in. The improved County CFT B was added in 1951 and following the launch of the New Fordson Major at that year's Royal Smithfield Show County introduced the Model Z agricultural crawler with the New Fordson Major diesel engine and its six forward speed and two reverse gearbox.

Launched in 1954 the County Four-Drive, also based on the New Fordson Major Diesel tractor, was the first County four-wheel drive tractor with equal-sized wheels. Heavy-duty roller chains from the rear axle drove the front wheels and the tractor was skid steered with a brake mechanism. Industrial applications included forestry work, sugar cane cultivation and site work with a dozer blade. A Fordson Power Major skid unit was used for the Mk II Four-Drive introduced in 1958 and a few Mk II Four-Drive tractors with a Fordson Super Major skid unit were built before it was superseded by the County Super-4 in 1961.

The County Model Z crawler was improved several times before the County Ploughman replaced it in 1957. The new Ploughman was also based on the New Fordson Major Diesel and the introduction of the Fordson Power Major resulted in the launch of the 51.8 hp County Ploughman P50 in 1958. The Ploughman 55 with the Super Major transmission and a 55 hp Ford industrial engine appeared in 1961 and the Clydesdale crawler, a no frills version of the P50, was added later that year.

County Commercial Cars were well established in the four-wheel drive tractor market in the early 1960s when they decided to concentrate on the production of wheeled tractors. The Clydesdale was discontinued in 1964 and the last P55 County Ploughman was made in 1965.

The County Super-4, introduced in 1961, was a conventional front-wheel steered tractor with a 52 hp Fordson Super Major engine and transmission, four equal-sized wheels and power-assisted steering. The front wheels were driven by two forward-running telescopic prop-shafts from the final reduction gears in the back axle. An optional rear-axle diff-lock gave a positive drive on all four wheels, the independent steering brake pedals operated both wheels on each side of the tractor and with the pedals locked the Super-4 had four-wheel braking for roadwork. Sales literature explained that the 'Super-4 reduces wheel spin, soil panning and smear and at the same time will reduce tyre wear, fuel consumption and running costs'.

The County Super-6 – with a 95 hp six-cylinder Ford industrial engine announced in 1962 – was very similar to the Super-4 and according to press publicity the Super-6 could do 'two hours work in sixty minutes on its four big equal-sized wheels that hardly know what wheel spin means'.

The 654 and 954 replaced the Super-4 and Super-6 when the Ford 1000 series tractors were launched in 1964 and County Commercial Cars used the new Basildon tractor range for the next ten years. The new model numbers denoted brake horsepower and the figure '4' indicated four-wheel drive. The 654 had a Ford 5000 engine and a six-cylinder 95 hp Ford industrial power unit was used for the 954. Both tractors had a Ford 5000 transmission and twin prop-shaft drive to the front wheels. Other equal-sized wheel County tractors

66. The County CFT with a Perkins engine was based on the E27N Fordson Major.

67. County entered the four-wheel drive market with the skid-steered Four-Drive in 1954.

based on the Ford 5000 series included the 854T, 1004 and 1124. A Ford 5000 engine was used for the 854T and the other two tractors had Ford industrial engines. Launched in 1967, the 854T was the first turbocharged County, the turbocharger being made by CAV. Model numbers still indicated engine power, the twin prop-shaft front-wheel drive system was retained and there was a choice of the Ford eight-speed manual gearbox or ten-speed Select-O-Speed transmission.

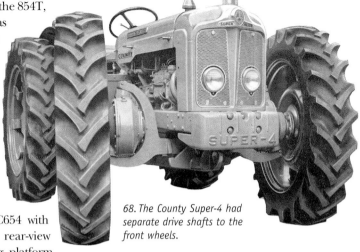

68. The County Super-4 had separate drive shafts to the front wheels.

Introduced in 1965 the County FC654 with front-mounted cab with full-width rear-view driving mirrors and rear load-carrying platform was the first of several County forward control tractors. Some were bought by agricultural contractors for spraying and fertiliser spreading but most were used for industrial and forestry work. The FC754, which replaced the FC654 when the Ford Force range appeared in 1968, was made for the next seven years. Four models of the FC1004 with six-cylinder Ford industrial engines were made between 1967 and 1977, the first three were based on Ford 5000 tractor skid-units and the last one – based on a Ford 6600 – appeared in 1975.

The County 4000-Four made between 1968 and 1975 was the first four-wheel drive County tractor with small front wheels. Select-O-Speed transmission was optional and apart from a transfer box and single prop-shaft drive to the front axle, the 4000-Four was virtually the same as the Ford 4000. County also introduced a four-wheel drive conversion kit in 1969 for the Ford 4000.

The restyled County 1164 had an integral cab, sloping bonnet and a fibreglass fuel tank. This model replaced the 1124 in 1971. It had the same six-cylinder engine as the American-built Ford 8000 and a strengthened 5000 transmission system. Later models of the 1164 had a Ford 7600 transmission and a Ford 8600 or Ford TW engine. The County 944, made between 1971 and 1975, with a turbocharged Ford 7000 engine and an eight forward and two reverse gearbox and Load Monitor had a Duncan cab. County still used the twin prop-shaft front axle drive system for the 1254 and 1454 launched in 1972. The 1254 with a Ford 8000 engine was made for three years while the

69. County forward control tractors were better suited to industrial use but this FC1004 was disc harrowing in a Suffolk field.

1454 with a Ford 9000 power unit was discontinued in 1978. Both tractors had a sixteen-speed Dual Power transmission and a Swedish-built cab. County Commercial Cars established a separate company called County Power Drives Ltd in the early 1970s to manufacture specialist models including the County 11F. Instead of a conventional clutch this two-wheel drive tractor, based on a Ford 7000, had a torque converter and an inching pedal for manoeuvring in confined spaces.

70. Dumper bodies and cranes were among the attachments used with the industrial version of the County FC1004.

71. The County 1164 had a Ford 8000 engine when it was launched in 1971 and a Ford 8600 power unit from 1975 until it was discontinued in 1977.

The Ford 600 and 700 series skid units were used for five different four-wheel drive County tractors with small front wheels. Made between 1975 and 1981 the 4600-Four, 6600-Four and 7600-Four were virtually identical to the two-wheel drive models with a single prop-shaft drive to the front axle. The 6700-Four and 7700-Four, added in 1978, were made until 1981.

Ford skid units were also used for the equal-sized four-wheel drive County 774, 974, 1174 and 1474 with flat floor cabs of the late 1970s. The County 774 with a Ford 6600 and later a 6610 engine and transmission and the 974 based on a Ford 7600 and later a 7610 were made well into the 1980s. Although most County tractors were based on Fordson and Ford skid units four-wheel drive conversions were also made for other tractors such as the International Harvester 614 and 634 and the Leyland 485 and 4100. Sales of County tractors had declined by 1980 when they introduced the improved 1884 based on the 188 hp Ford TW30. A TW35 skid unit was used in later years and both were similar to the 1474 with a sixteen forward and four reverse speed Dual Power transmission.

County Commercial Cars went into receivership in 1983 and a County tractor dealer at Ludlow purchased the business. The company was renamed County Tractors Ltd and production of equal-sized four-wheel drive tractors continued at Fleet in Hampshire. Fourteen different models were made there in 1984. Twelve of them, from the 80 hp 764P to the 195 hp 1884, were four-wheel drive models while the high-clearance 762H and the two-wheel drive County 1162 completed the range. The Benson Group bought County Tractors Ltd in 1987 and moved to Knighton in Wales where production of the 774, 974, 1184 tractors and the County Hi-Drive conversion kit was scaled down. The Hi-Drive kit, with a high front axle, drop housings for the rear axle and extended cab steps with an optional lower hydraulic linkage and drawbar, was suitable for Ford 5610, 6610 and 7610 tractors. The last County tractor was made at Knighton in 1990.

Having received permission to use the County name, SEM Engineering at Basildon introduced the County HSH140 in 1994. Based on the Ford New Holland 40 series this high-speed haulage tractor designed for worldwide use had a top speed of 25 mph. The specification included a six-cylinder 143 hp Powerstar engine with a large capacity radiator and an additional engine cooler for tropical conditions and the choice of an eight forward and two reverse gearbox or six-speed

72. The 4600-Four was the first four-wheel drive County tractor with small front wheels.

73. The County 1474 with a 149 hp turbocharged engine was launched at the 1978 Royal Smithfield Show.

power shift transmission with a torque converter. Other features of the HSH140 included an industrial front axle, coil spring suspension, shock absorbers, a diff-lock, hydrostatic steering, three-point linkage and an air-over-hydraulic braking system with a built-in compressor. SEM Engineering was also involved in the development of some new prototype County tractors, mainly for overseas use, based on the Ford 40 series but within a year the company had ceased trading.

Crawley

The 7 hp Crawley 75, introduced in 1958, was one of a number of ride-on tractors for smallholdings and market gardens in the 1950s and 1960s. Made by Crawley Metal Productions at East Preston near Littlehampton with some parts sourced from Holland most Crawley tractors had a 7 hp Petter PC1 Mk II air-cooled diesel engine while others were sold with an 8 hp JAP Model 5 petrol or vaporising oil engine. The Petter diesel engine used about 1½ gallons of fuel in a ten-hour day compared with a daily consumption of two gallons by the petrol and vaporising oil engines.

The Crawley 75 had an automatic centrifugal clutch, three forward gears and one reverse, a worm and wheel final drive, independent rear wheel brakes, draft control hydraulics with a belt-driven hydraulic pump, a centre power take-off shaft and universal drawbar. The wheel track was

74. SEM Engineering made the 143 hp County HSH 140 at Basildon.

75. The Crawley 75 made by Crawley Metal Productions was designed in America.

adjustable in four steps from 32 to 48 in achieved by repositioning one or both of the wheel discs on their hubs. Optional extras included a foot throttle and rear power take-off shaft. The diesel-engined Crawley 75, which cost £290 in 1958, was advertised as the first tractor of its size able to stand up to heavy work. It was also suggested that every farm should have a Crawley as it could do every job on the farm including ploughing, drilling, spraying, spreading fertiliser and haymaking with a mid-mounted mower. A further incentive to buy a Crawley was the offer of two free services during the first eighteen months of its life.

David Brown

A small factory established by David Brown in Huddersfield in 1860 was an important gear wheel manufacturing business by the early 1900s. The founder's grandson, also called David Brown, joined forces with Harry Ferguson in 1936 to manufacture the Ferguson-Brown Type A tractor with the revolutionary Ferguson draft control hydraulic system.

A 20 hp petrol or paraffin Coventry-Climax engine was used for the first Ferguson-Brown tractors but from 1937 a similar David Brown engine was used for the Type A tractor. It had a three forward and one reverse gearbox with a top speed of just under 5 mph, spade lug wheels and independent drum brakes with an optional power take-off and side-mounted belt pulley. The steel-wheeled Ferguson-Brown with a petrol engine cost £198 in 1938, the paraffin tractor was an extra £10 and pneumatic tyres, available from 1937, added another £40 to the bill. Many farmers considered the Ferguson-Brown too expensive when compared with the £100 price tag for a Fordson Model N. Approximately 1,300 Ferguson-Browns had been made at Huddersfield by 1939 when the relationship between the two tractor men came to an end and production of Ferguson tractors was transferred to America.

Plans to build the David Brown VAK1 (Vehicle Agricultural Kerosene) and VAG1 (Vehicle Agricultural Gasoline) were well advanced when Harry Ferguson and David Brown parted company and the new tractor was launched at the 100th Royal Show at Windsor in 1939. The 35 hp VAK1 had an overhead-valve four-cylinder,

76. Hunting pink was the standard colour for early David Brown tractors including the VAK1C made between 1947 and 1953.

water-cooled petrol or paraffin engine, a four forward speed and one reverse gearbox and independent brakes. Power take-off was optional but hydraulic linkage, which required wheels to control the implement, was not available until 1941. Early VAK1 tractors had a tubular front axle and a rather fragile cast-iron radiator grille but later models had a perforated sheet steel radiator grille. The VIG1 (Vehicle Industrial Gasoline) was an industrial version of the VAG1 made Power take-off was optional but hydraulic linkage, which required wheels to control the implement, was not available until 1941.period and many of them were used by the Royal Air Force to tow aircraft and bomb trolleys.

The VAK1A and VAG1A, introduced in 1945, had longer wheelbases, square section front axles and patented adjustable top links. A modified manifold gave a faster engine warm up and by lowering the position of the carburettor David Brown was able to dispense with the fuel pump. The VAK1C with a four forward and one reverse gearbox, which replaced the VAK1A in 1947, was the first David Brown Cropmaster tractor. From 1949 it was also the first British tractor to have a six-speed gearbox and two-speed power take-off. A pedal and a hand lever were provided to control the clutch, other features included hydraulic linkage, swinging drawbar and electric lighting. David Brown made their own range of mounted ploughs and

cultivating implements and various other machines including Allman sprayers, Robot planters and Byron disc harrows were officially approved for use with the Cropmaster tractor. An optional 34 hp diesel engine introduced for the VAK1C Cropmaster in 1949 was used until the tractor was discontinued in 1953. Variants of the VAK1C included a 54 in wide vineyard model, the VTK1 Thresherman and the V1G1AR Taskmaster for airport and industrial use. The Super Cropmaster – with a more powerful 38 hp vaporising oil engine, 12-volt electrics, side panels on the engine bonnet and wider tyres – was added to the range of David Brown tractors in 1950.

The first David Brown crawler tractors appeared in 1942 when they launched the DB4 with a 38 hp Dorman diesel engine and a five-speed gearbox. The Trackmaster 30, a crawler version of the Cropmaster with differential steering and a six-speed gearbox, which replaced the DB4 in 1950, was made for three years. The petrol/vaporising oil engine crawler was initially known as the TAK3 (Tracklayer Agricultural Kerosene) and the diesel-engined model was badged as the TAD3 (Tracklayer Agricultural Diesel). Another new crawler, the Trackmaster Diesel 50 with a six-cylinder engine and six-speed gearbox was added in 1952. When David Brown introduced the 25, 25D, 30C and 30D in 1953 the crawlers were renumbered to match the new wheeled

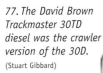

77. The David Brown Trackmaster 30TD diesel was the crawler version of the 30D.
(Stuart Gibbard)

tractors. The Trackmaster 30 and 30 Diesel became the 30T and 30TD and the Trackmaster 50 Diesel was badged as the 50TD.

The David Brown 25 and 30 tractors with a single pan seat, made between 1953 and 1958, retained some features of the earlier Cropmaster but they did not have a scuttle or wide mudguards. The David Brown 25 had a 32 hp vaporising oil engine or a 37 hp petrol engine while the 25D, launched in late 1953, had a 31 hp diesel engine. The 30C and 30D had petrol, vaporising oil or diesel engines rated at 41, 37 and 34 hp respectively, making them similar to the David Brown 25. The 30D had a decompressor on all cylinders for cold starting. The standard specification included a six-speed two-range gearbox with a top speed of 15 mph, a two-speed power take-off and belt pulley, hydraulic lift, independent expanding shoe brakes and a parking brake. The clutch was operated with the usual foot pedal or an alternative hand lever at the back which gave the driver 'full control of the tractor when standing at the rear'. An optional hydraulic overload mechanism in the top link, connected by a cable to the clutch hand lever, automatically disengaged the clutch if the implement hit an obstruction. The Ferguson draft control hydraulic linkage was still protected by patents so David Brown, like other manufacturers, developed their own weight transfer system. The Traction Control Unit (TCU), introduced in 1953, required depth wheels on the implement. This meant that wheelslip could be eliminated by varying the amount of the implement weight transferred on to the tractor. TCU could also be used with a special linkage to transfer some of the weight of trailed equipment on to the tractor.

Launched in 1953 the wheeled David Brown 50D and 50TD crawler – no longer with the Trackmaster logo – had the same 50 hp six-cylinder diesel engine but the 50D could only manage a sustained drawbar pull of 6,550 lb compared with the 9,800 lb pull at the 50TD's drawbar. The 50D, still with a bench seat, was primarily a towing tractor with a four-speed power take-off shaft but there was no hydraulic linkage. The rear wheels could be moved in or out on their axles for easy wheel track adjustment and some farmer/contractors used the 50D with its optional side-mounted belt pulley to drive a threshing machine.

78. David Brown 30 and 30D tractors had a six-speed gearbox, two-speed belt pulley and two-speed power take-off.

79. The 50D was the only David Brown tractor with a side-mounted belt pulley used by some agricultural contractors to drive their threshing tackle.

Harrison, McGregor and Guest had been manufacturers of Albion binders, mowers and other farm machinery at Leigh in Lancashire since the 1870s. The business was acquired by David Brown Tractors in 1955. This enabled the Meltham company to add the David Brown Albion trailed combine harvester, the Hurricane forage harvester, a pick-up baler and other implements to their product range.

David Brown launched the 14 hp 2D tool carrier at the 1955 Royal Smithfield Show. The specification included a twin-cylinder four-stroke, air-cooled, diesel engine, a single-plate clutch and a four forward and one reverse gearbox all mounted above the rear axle. Independent drum brakes, adjustable wheel track and the Air-Light compressed air implement lift system were standard. A small compressor on the gearbox provided compressed air for two lift cylinders used to raise and lower the underslung and rear-mounted toolbars. The two control levers could be locked together or used separately for independent control of each lift cylinder, for example when using a reversible plough. The tubular steel chassis served as an air reservoir and a connection for a tyre inflator was an added bonus. Although claimed to be a maid of all

work the David Brown 2D was more at home working in rowcrops with its 8 ft or 10 ft wide mid-mounted toolbar. Variants included the narrow 3 ft 3 in wide 2D vineyard model and a shortened version for industrial use.

The 2D with electric starting cost £399 in 1956 when sales catalogues listed five standard 2D tractor and implement packs for dairy farms, commercial growers, rowcrops, smallholders and market gardeners. The dairy farm pack, for example, consisted of a tractor, a 5 ft mid-mounted cutter bar mower, a trailer and front axle weights. The commercial grower pack, which cost £514, included a tractor, a toolbar with depth control wheels, markers, ridging bodies and hoes together with a single-furrow one-way plough with disc coulters, skimmers and depth wheels.

The distinctive hunting pink David Brown 900 with blue wheels and radiator grille was launched at the 1956 Royal Smithfield Show. It had six forward and two reverse gears and there was the choice of a 40 hp diesel, a 37 hp vaporising oil or a 45 hp petrol engine. As with earlier David Brown tractors the steering column was offset to the right. The 900 was one of the first diesel-engined tractors to have a distributor-type

80. The rear-engined David Brown 2D made between 1956 and 1961 was designed for smallholders and market gardeners.

fuel injection pump and mechanical governor. An improved model with a live two-speed power take-off, a central steering column and dual category TCU hydraulics appeared in 1957. A few tricycle-wheeled and high-clearance David Brown 900 tractors were made mainly for the American market. They could be used with an underslung toolbar.

The 900 was not the most successful David Brown tractor and in 1958 it was replaced by the 950. The new model's yellow wheels and radiator grille distinguished it from the 900 but otherwise it was much the same at its predecessor. The main changes were an improved diesel fuel injection system that added an extra 2.5 hp, a modified drawbar and redesigned steering linkage. David Brown claimed that the improved 950 Implematic tractor – the first of a new family of Implematics, which replaced the earlier model in 1959 – overcame the need to buy a new tractor compatible with existing implements on the farm. Sales literature explained that the 950 Implematic would give the best possible results no matter whether it was being used with the TCU system for mounted implements fitted with depth wheels or the new Implematic Traction Depth Control

system for Ferguson-type implements without depth wheels. The Traction Depth Control system transferred variations in the force acting through the top link by means of a Bowden cable from the tractor end of the top link to the hydraulic control valve. Any change in this force automatically raised or lowered the implement to maintain the pre-selected draft setting and, in ideal conditions, a constant depth.

David Brown launched the 850 Implematic in 1960 followed by the 880 and 990 Implematic models in 1961. The standard and narrow width 35 hp 850 and 42.5 hp 880 Implematic tractors had six forward and two reverse gears and a diff-lock with optional Live-Drive power take-off and hydraulics. The 880 had category I and II hydraulic linkage but the 850 was restricted to category I implements. Tractor electrical systems had a positive earth at the time but the 880 was the first tractor with a negative earth electrical system to comply with a new European standard. A 35 hp petrol engine, available for the 850, was only made for a short while. A number of David Brown 850 and 950 tractors, painted green and white, were sold in America by the Oliver Corporation in the early 1960s badged as the Oliver 500 and Oliver 600.

The 52 hp 990 Implematic announced at the 1961 Royal Smithfield Show was the first David Brown tractor with a cross-flow cylinder head. It had a dual range six forward and two reverse gearbox, a choice of a Live-Drive or standard transmission and an optional automatic gearbox that changed gear according to the load on the tractor. However, farmers were not willing to pay an extra £250 for the automatic gearbox and the idea was dropped. An improved 990, announced in 1963, had the battery repositioned in front of the radiator, a strengthened front axle and an optional twelve-speed gearbox.

81. The David Brown 950 was sold in America as the Oliver 600.

Launched in 1964 the three-cylinder 33 hp 770 with a two-range twelve forward and four reverse gearbox was the first David Brown tractor with the new Selectamatic hydraulic system. The driver selected depth control, TCU, height control, or external services on the Selectamatic dial and then controlled the chosen system with the hydraulic lever.

A chocolate brown and orchid white colour scheme with a poppy red silencer, suggested by potential David Brown tractor dealers in America, replaced the familiar hunting pink and yellow paintwork

82. Later, and much improved, models of the David Brown 880 had a chocolate brown and orchid white colour scheme.

83. The chocolate brown and orchid white colour scheme replaced the red and yellow paintwork in 1965. This David Brown 990 is hitched to a Vicon Rotaspa spading machine.

in October 1965 when the new paintwork was also used on the home market. The 880 and the 990 Implematic hydraulic systems were superseded by the new David Brown Selectamatic hydraulics with an engine-mounted pump in 1965 and the power ratings for the 770 and 990 tractors were increased to 36 and 55 hp in the same year.

David Brown had an annual output of about 30,000 tractors when the 67 hp 1200 Selectamatic was launched in 1967 to meet the demand for more horsepower. The 1200 had a four-cylinder direct-injection engine, six forward and two reverse gears with a top speed of 14.7 mph and an optional twelve forward speed gearbox.

84. A crankshaft-driven hydraulic pump provided the oil for the 67 hp David Brown 1200 Selectamatic's live hydraulic system.

85. The David Brown 1412 had a 91 hp turbocharged four-cylinder engine, a dry air cleaner and oil-immersed disc brakes.

Other features included an independent power take-off, diff-lock, drum brakes and live Selectamatic hydraulics with an engine-driven pump, power steering and a pedal-operated exhaust brake. This device was a form of pneumatic engine brake which, when a pedal was depressed, restricted the flow of exhaust gases from the engine and slowed the transmission and rear wheels. The 1200's engine power was increased to 72 hp in 1968 and a four-wheel drive model with a Selene front axle was added in 1970. The David Brown 1200 Selectamatic appeared in 1967. It was the first of six new Selectamatic tractors, including the 885, 990, 995, 996, 1210 and four-wheel drive 1212, with synchromesh gearboxes under their brown and orchid white paintwork. The three-cylinder 48 hp David Brown 885 with a synchromesh gearbox, which replaced the 880, was the smallest model in the range and the 990 had 58 hp under the bonnet. The 995 and 996 were rated at 64 hp. Other David Brown tractors at the time had dual clutches, while the 996 had a separate hand clutch for the live power-take off. A

Selene front axle was used for the 72 hp four-wheel David Brown 1210 with category II hydraulic linkage. An earlier attempt to market an automatic David Brown gearbox in 1961 failed but the semi-automatic Hydra Shift transmission introduced ten years later was an award winner. The 72 hp 1212 Hydra Shift tractor, which replaced the 1200, had twelve forward and four reverse speeds provided by four clutchless change-on-the-move gears in each of three forward ratios and one in reverse.

Tenneco, which already owned Case, completed their acquisition of David Brown Tractors in 1972 and the Case influence became apparent in the UK in 1973 with the introduction of the orange and white six-cylinder 101 bhp Case 970 Agri-King. The synchromesh David Brown Case 1410 and Hydrashift 1412 with turbocharged 91 hp engines, twelve forward and four reverse gears, live hydraulics, independent power take-off and disc brakes were launched in 1975. Change-on-the-move four-wheel drive models of the 1212 and 1412 were added in 1976. The 500,000th David Brown tractor was due to come off the production line in 1977 when the Queen celebrated her Silver Jubilee. To mark the occasion David Brown produced a specially painted tractor with a silver cab and purple roof which was sold that year by auction at the Royal Smithfield Show.

The 90 Series tractors with 'David Brown' on the bonnet sides and 'Case' on the radiator grille appeared in 1979. Launched two years earlier in America the Case 90 Series included five 48–103 hp models made at Meltham and seven 120–273 hp Case tractors built in America.

The two-wheel drive 1190, the smallest of the Meltham-built Series 90 tractors, had a twelve forward and four reverse speed synchromesh gearbox and hydrostatic steering. The 1290 and 1390 were rated at 58 and 67 hp respectively and had the same basic specification as the 1190. They also had the option of four-wheel drive and a high clearance conversion kit. An optional Hydra Shift transmission for the 1390 was added in 1983. The 83 hp 1490 was the biggest four-cylinder engine 90 Series tractor and the 103 hp 1690 with a six-cylinder engine was the flagship model in the Meltham range of David Brown Case tractors. Features of the four-wheel drive 1690 included a 540/1,000 rpm independent power take-off and lower link sensing hydraulics. The American-built four-wheel drive Case 4690 and 4890 with equal-sized wheels and a twelve-speed powershift transmission were introduced to British farmers at the 1981 Royal Show. The David Brown name disappeared from the UK market when the new Case 94 series was launched in 1983.

Deutz

Nikolaus August Otto, who perfected the four-stroke or Otto cycle engine in 1876, founded an engineering company with Eigen Langen at Deutz near Cologne in 1864. The first Deutz farm tractor with a 26 hp petrol engine with two large flywheels and four equally large steel wheels was built by a subsidiary Deutz company at Philadelphia in America in 1894. A 25 hp Deutz tractor with a plough at each end was made in Germany in 1907. It had two seats and a central steering wheel so that the driver was able to change position after each run to get a clear view of the plough as it worked its way back and forth across the field. Another 1919 vintage Deutz tractor for farming and forestry work looked more like a short wheelbase lorry than a 40 hp farm tractor.

Dr Rudolph Diesel, who worked at Deutz in the 1890s, made the world's first high compression self-igniting (diesel) engine in 1897. An improved version of this power unit was used for the first diesel-engined 14 hp Deutz MTH 222 tractor made in 1927. A 25–28 hp water-cooled diesel-engined Deutz tractor with two vertical cylinders, a five-speed gearbox, power take-off and belt pulley appeared in 1933 and the 11 hp Deutz Bauernschlepper or smallholder's tractor was introduced in 1936. The Bauernschlepper, the world's first small tractor to be manufactured in large numbers, played a major role in the mechanisation of small farms in Germany.

There were many mergers in the German farm machinery industry in the 1930s. They included the formation of Klöckner-Humboldt-Deutz AG (K-H-D) in 1938 and within four years they were mass-producing Deutz water-cooled diesel engines. The F1L 514 tractor with a 15 hp single-cylinder engine and four forward gears introduced in 1949 was the first of many Deutz tractors with air-cooled diesel engines. Early 1950s tractors included the 11 hp F1L 612 rowcrop model, wheeled and crawler versions of the 28 hp F2L 514 with five forward gears, the 50 hp F3L 514, the F4L 514 with a 60 hp engine and seven forward speeds. The 100 hp six-cylinder Deutz DK 100 crawler completed the range. The figure after the letter F in the tractor model number (e.g. F4L) indicated the number of engine cylinders.

The new Deutz FL 712 series tractors, including the F1L 712/D-15 and F2L 712/D-25, launched in 1959 and by the early 1960s the K-H-D range included twenty-five wheeled tractors up to 65 hp and two tracklayers with 75 and 100 hp air-cooled engines. Independent front axle suspension was a special feature of the 15 hp D-15 with six forward and two reverse gears and a diff-lock. The 22 hp twin-cylinder D-25 tractor had an eight forward and two reverse gearbox and the new Deutz Transfermatic three-point linkage hydraulic system was optional on both models.

A sales leaflet of the time explained that the 'D-15 is ideally suited for small farms of 25–30 acres or as a second tractor on large farms for tilling, mowing and haulage work'. It was also suggested that 'if the purchase of the lift linkage can not be immediately afforded it will be found advantageous to decide on purchasing the three-point linkage system with a

86. Independent suspension on the front axle was a feature of the 15 hp Deutz D-15.

87. The air-cooled three-cylinder 38 hp Deutz D-40S had seven forward and three reverse gears, ground and engine speed power take-off and a diff-lock.

manual lifting device. It will then be found easier to switch to the hydraulic power lift at a later stage.'

The 28 hp air-cooled D-30 and D-30S with eight forward speeds replaced the D-25 when British Deutz Ltd opened an office on the Strand in London in 1961. Advertisements explaining the benefits of air-cooled engines pointed out that there was no laborious nightly radiator draining when the tractor was left outside in frosty weather. This type of engine was popular with continental farmers and the Deutz engine design with many interchangeable parts made it possible to build one-, two-, three-, four- or six-cylinder in-line and vee-engines with up to twelve cylinders on the same production line.

The Transfermatic hydraulic lift was introduced in 1962 when the Deutz range included seven wheeled models from 15 to 65 hp and two crawlers for farm and industrial use with 75 and 100 hp engines. The D-40S, D-50S and D-80S standard wheeled tractors with three-, four- and six-cylinder engines were added during the year. The two-cylinder 28 hp rear-engined Multitrac, introduced in 1962, was

the first of many Deutz tool carriers. The 39 hp Allrad (four-wheel drive) with four equal-sized wheels and a front-mounted engine was another early Deutz tool carrier.

K-H-D became part owners of Fahr Machinenfabrik in 1968 and the acquisition of this well-established German tractor and farm machinery manufacturer, which also made a range of implements for Deutz tractors, was completed late in 1970. Watveare Overseas Ltd, derived from Watkins & Roseveare, had premises at Ivybridge in Devon and Westbury in Wiltshire and were appointed UK distributors in 1974 for the pea-green Deutz 06 series tractors with air-cooled direct-injection engines. The two- and four-wheel drive D7206/D7206A and D8006/D906A tractors – first seen in Britain early in 1975 – had four cylinders and a twelve forward and four reverse gearbox and the four-wheel drive six-cylinder D100 06A had sixteen forward and seven reverse gears. The model number indicated horsepower and the letter A denoted four-wheel drive. Hydrostatic steering and the Deutz Transfermatic hydraulic system were standard across the range.

Sales literature described Transfermatic hydraulics as a system with 'joltless control and an even working depth' and 'single lever operation after selecting either draft or position control'. The raised implement was 'carried on a cushion of oil in the service cylinder to provide a shock absorber effect so that the tractor and implement were subject to less fatigue'. The 06 series

with various improvements and additions, including the D4506 with a three-cylinder Deutz engine and an eight forward and four reverse gearbox, was made until the early 1980s. Later 06 series tractors had a safety cab.

A new generation of Deutz self-propelled tool carriers was launched in 1972 and was introduced to UK farmers in 1975. The two-wheel drive Intrac 2003 and the four-wheel drive 2003A with a glass-walled cab and the driving seat in front of the 62 hp air-cooled diesel engine had an eight forward and four reverse synchro-mesh gearbox with an optional creeper box adding four extra speeds from 0.2 to 1 mph. The standard specification included power-assisted steering, front and rear independent power take-off shafts and live hydraulic linkages with top link sensing. The driver was also seated at the front of the Intrac 2004 tool drive carrier imported in the early 1980s by Watveare Overseas at Westbury in Wiltshire. The four-wheel Intrac 2004 had an air-cooled 70 hp Deutz diesel engine and a twelve forward and four reverse gearbox with a top speed of just under 19 mph.

The 80 hp DX85 and 102 hp DX110/A that took part in the 1979 Long Sutton tractor tests were early models in the new Deutz DX range. The range included the two-

88. The model number of the mid-1970s Deutz D 130–06 meant that it was a four-wheel drive tractor with a six-cylinder 130 hp engine.

wheel drive DX110, the 52 hp D5206 and the four-wheel drive DX90A with a five-cylinder 88 hp engine, fifteen forward and five reverse gears and an offset drive shaft to the front wheels. K-H-D adopted the Deutz-Fahr name for its agricultural equipment in 1981 and the four-wheel drive Deutz-Fahr DX120A and DX145A were launched in the same year. The new tractors had 110 and 132 hp engines with thermostatically controlled cooling fans claimed to reduce fuel consumption by 10 per cent and a twenty-four forward and eight reverse Powermatic change-on-the-move transmission.

The first Deutz-Fahr 07C (C stood for comfort) medium-power tractors in the 35–75 hp bracket replaced the 06 series in 1980. With the exception of the 35 hp tractor the 07 range with an improved cab and easier servicing was available with four-wheel drive. The smaller models had nine forward and four reverse synchromesh transmissions and telescopic front axles while the other tractors were equipped with a twelve forward and four reverse gearbox. There were seventeen two- and four-wheel drive Deutz-Fahr models on the UK market in 1983, including the 35–82 hp 07C and the 78–220 hp DX.

The new 78–145 hp four- and six-cylinder DX4 and DX6 four-wheel drive tractors that were introduced at the 1983 Royal Smithfield Show had a central drive shaft to the front axle, independent front hydraulic linkage and improved ground clearance. The DX3 models with three-cylinder 50–82 hp engines superseded the 07C series in 1984 and the launch of the flagship 220 hp eight-cylinder DX8.30 with Powermatic transmission and electronic linkage control hydraulics completed the DX range. An advertisement for four-wheel drive Deutz DX6 tractors explained that they were 'not only more powerful in appearance, they've got the muscles you've always wanted including hydraulics that will outlift most of the competition'. The DX6 range with 110–145 hp air-cooled Deutz diesel engines had a fifteen forward and five reverse synchromesh gearbox and the Optitrac diff-lock on the front axle.

89. The 1970s 73 hp Intrac 2004/2004A with a twelve forward and four reverse gearbox had the Deutz Transfermatic hydraulic linkage.

K-H-D bought the engine manufacturer Motoren-Werke Mannheim (MWM) in 1985 when they launched six new 50–98 hp Deutz DX3 and DX4 HPE (High Performance and Economy) tractors. Features of the HPE models included a synchromesh gearbox and a new cab, which according to sales literature brought a new standard of luxury to the 50–100 hp tractor market. K-H-D also bought Allis-Chalmers in 1985 when they were making the five-cylinder 95 hp Fiat-Allis 9130 and the six-cylinder 120 hp Fiat-Allis 9150 with air-cooled Deutz engines for the American market. It was all change again in 1990 when AGCO bought Fiat-Allis and the tractors were sold in America under the AGCO-Allis brand name.

Deutz-Fahr celebrated sixty years of tractor production with the Diamond range introduced to UK farmers at the 1986 Royal Agricultural Show. The Diamond tractors had a multi-speed transmission, remote hydraulic linkage controls, additional cab soundproofing and a stereo radio cassette for driver entertainment. The optional Deutz Agtronic performance monitoring system also introduced in 1986 for tractors over 100 hp could be used to record data on forward speed, area covered and power shaft speed, among others.

The 72–117 hp DX Commander series launched in 1989 with the latest Command Centre cab had an improved four-wheel drive, four-wheel braking and OptiTrac self-locking differential on the front axle which disengaged automatically when the front wheels were turned to the left or right. A growing interest in the late 1980s in the use of bio-fuels for diesel engines

90. The Deutz DX3 Series launched in 1983 had three-cylinder engines in the 50–82 hp bracket.

prompted Deutz-Fahr to approve the use of rapeseed oil for all Deutz tractors made after 1989. However, it was pointed out that although power output would not be affected the engine would require more frequent servicing.

Five new Deutz-Fahr Agrostar DX4 and DX6 four- and six-cylinder 88–143 hp tractors were launched in 1990. They had forty-eight forward and twelve reverse synchromesh transmissions and the new suspended flat floor cab with the latest Agrotronic performance monitor was advertised as the quietest cab on the market.

K-H-D introduced the first high visibility Deutz AgroXtra tractors in 1991 and with various other changes the 1992 Deutz-Fahr range consisted of the AgroPrima, AgroStar and AgroXtra tractors. The high-visibility AgroXtra 4.17A, 4.47A 4.57A, 6.07A and 6.17A tractors with sloping bonnets had 80–113 hp air-cooled Deutz engines and there was a choice of transmissions with up to twenty-four forward and eight reverse speeds and hydraulic self-adjusting disc brakes.

The computerised AgroTronic HD system designed to combat the see-sawing effect caused when driving across rutted ground was an optional extra on the Agro Extra Power Plus models launched in 1993. The ASM drive control management system for the front and rear diff-locks and front-wheel drive, designed to give the

91. The six-cylinder Agrotron 230 and 260 had Deutz engines, a forty forward and reverse powershift transmission and an air-suspension system for the cab.

driver maximum comfort and avoid operating errors, was a feature of the new 107, 115 and 125 hp AgroStar 6.08A, 6.28A and 6.38A tractors launched in 1994.

K-H-D sold the Deutz-Fahr tractor business to SAME Trattori in 1995 and this led to the formation of the SAME Deutz-Fahr group. Watveare, who had moved to Warwickshire in 1990, retained the Deutz-Fahr distributorship until 1997 when SDF established a marketing organisation in the UK.

The Agrotron series initially launched in 1996 was extended to ten models in the 75–150 hp bracket in 1997 and even more powerful Agrotrons with 230 and 260 hp engines appeared in 1998. The most significant change was from air-cooled power units to water-cooled, low emission engines. Other features included load-sensing hydraulics, a front-axle suspension system and high visibility air-conditioned cabs bristling with the latest electronic technology. Deutz Agrokid, Agroplus and Agrotron four-wheel drive tractors were included in the SAME-Deutz-Fahr price list for 2000. The three Deutz Agrokid compact tractors with 25, 35 and 42 hp Deutz water-cooled engines had a twelve forward and reverse synchromesh gearbox, front, centre and rear power take-off shafts and the option of a safety frame or quiet cab. An optional three-range hydrostatic transmission was available for the 25 hp Agrokid.

A continuously variable transmission (CVT) was a new feature on the 150 hp Deutz-Fahr Agrotron TTV1160 introduced to British farmers at the 2002 Cereals event. The new tractor with an electronically managed six-cylinder turbocharged engine had front axle suspension and a suspended cab. The CVT transmission, with infinitely variable speeds of up to 50 kmh, was split into four working ranges operated with a joystick on the right-hand armrest of the driving seat. The 2005 Deutz-Fahr tractor range included five Agroplus and twelve Agrotron models. The air-cooled Agroplus engines ranged from the three-cylinder 60 hp Agroplus 60 to the six-cylinder Agroplus 100 with 101 hp under the bonnet. The 60 and 77 hp models had a twenty forward and ten reverse gearbox and the 76–101 hp tractors were available with a synchromesh or powershift transmission. Agrotron tractors had water-cooled four- and six-cylinder intercooled diesel engines with maximum outputs in the 100–262 hp bracket. A powershift transmission was standard and with the exception of 100

and 110 hp tractors which had twenty-four forward and eight reverse gearboxes, the other Agrotrons had transmissions with forty forward and reverse speeds.

The 90–110 hp K series Agrotron tractors introduced in 2005 had glass-roofed cabs to improve visibility for front load work and there were no less than seventeen vents for the air-conditioning and heating system. The tractors had a four-speed power take-off with 540 and 1,000 rpm operating speeds at 1,900 rated engine speed and at the 1,550 rpm economy setting.

Doe

Farmers in Essex and surrounding areas bought Allis-Chalmers, Case and Fordson tractors from Ernest Doe in the 1930s. Ernest Doe & Sons Ltd at Ulting in Essex were leading Fordson tractor dealers when they introduced the Doe Dual Power in 1958 to meet a request from some heavy land farmers for a more powerful wheeled tractor to replace their crawler tractor for heavy cultivation work.

An Essex farmer came up with the idea of linking two Fordson Power Majors in tandem to make a 100 hp four-wheel drive tractor. The front axle was removed from both tractors and the rear unit was close-coupled on a 2 ft diameter steel turntable attached to a strengthened drawbar on the leading tractor. An engine-mounted hydraulic pump supplied oil to the steering rams used to turn the front tractor through an angle close to 90 degrees. The driver, seated on the rear unit, controlled both tractor clutches with a hydraulic master cylinder and slave rams. A linkage enabled the driver to select high or low range on the front tractor while seated on the rear unit but it was necessary to dismount in order to select the required forward gear on the leading tractor. After returning to the seat the driver set the Dual Power in motion by engaging the range gearbox on the front tractor and then releasing the front and rear unit clutch pedals. The tractor was reversed with only the rear tractor in gear. The Doe Dual Power was supplied on pneumatic tyres or spade lug wheels. About six tractors had been made when an improved Doe Dual Drive with an assistor ram for the three-point linkage and improved steering was introduced in 1959.

The 1960 Triple-D, based on two 52 hp Fordson Super Major tractors, had a more efficient method of changing

92. The Fritzmeier cab on this Doe Triple-D with a Doe heavy-duty cultivator provided some protection from the weather.

gear using a hydraulic master cylinder and slave rams meaning that the driver was able to change gear on both units without leaving the seat. The front diff-lock was also engaged hydraulically from the rear tractor. The Triple-D had 108 hp under its two bonnets when the new performance 54 hp Super Major appeared in 1963. Ernest Doe also made a number of Super-Power implements for the Triple-D including three- and four-furrow reversible ploughs and six-, seven- and eight-furrow conventional semi-mounted steerable ploughs,

subsoilers and cultivators. The Doe implement range also included a 14 ft wide, fifteen-tine ripper cultivator and a hydraulically folded 20 ft toolbar with two gangs of disc harrows or spring tine cultivator tines.

When Ford launched the 6X series in 1965 two Ford 5000 tractors with a combined 130 hp were used for the Doe 130. The clutch, gear changing and steering were hydraulically operated and a cable connected the throttle levers. An improved Ford 5000 with a 75 hp engine appeared in 1968 and a few of them were used to

93. Three hundred and fifty Doe Triple-D, Doe 130 and Doe 150 tractors were made between 1958 and 1965.

94. The 130 hp Doe 130 with its strengthened hydraulic system to cope with the increased drawbar horsepower cost £3,300 in 1967.

95. The Doe 5100 was a stretched version of the Ford 5000 with a 100 hp six-cylinder engine.

build the Doe 150. About 350 Doe Triple-Ds had been made at Ulting by 1970 when production came to an end partly due to competition from County, Roadless, Muir Hill and other four-wheel drive tractors and the high development costs involved in complying with the new safety cab regulations.

Ernest Doe & Sons briefly returned to tractor production in 1971 when they made the two-wheel drive Doe 5100 which was a stretched and strengthened Ford 5000 with a 100 hp six-cylinder Ford industrial engine, a heavy-duty clutch, power-assisted steering and wet disc brakes. The tractor was seen at several public demonstrations but very few were made.

Dutra

The Dutra story began in 1912 when HSCS (Hofherr-Schrantz-Clayton-Shuttleworth) was established at Budapest in Hungary to manufacture steam engines. HSCS built their first tractors in 1923 with a petrol, paraffin or producer gas single-cylinder horizontal engine and the first HSCS tractors with 15–30 hp hot-bulb semi-diesel engines appeared in 1924. The HSCS R-30-35 series of wheeled and crawler tractors with hot bulb engines and three-speed gearboxes were made from 1936 until the early 1940s. Lanz owned the Budapest factory from 1938 until 1951 when it was nationalised by the Hungarian State and became the

96. Dutra tractors were on sale in the UK between the mid-1960s and the early 1970s.

Red Star Tractor factory. Production of tractors with single-cylinder semi-diesel engines continued there until 1956. A twin-cylinder tractor similar to the John Deere Model D was produced at the nearby Csepel Engineering factory from 1930 until the early 1940s when military vehicles were made there. Nationalised at the end of the war, the Csepel factory made diesel engines and trucks under licence from Steyr-Daimler-Puch. Dumper trucks and four-cylinder 50 hp wheeled and crawler tractors were added in the early 1950s when the Dutra brand name, a combination of DUmper and TRActor, came into use.

A new range of 28 hp UE-28 wheeled tractors with twin-cylinder Csepel engines was introduced in 1956 followed by an equal-sized wheel 28 hp four-wheel drive model. The 40 hp UE-40 was added in 1960 and the four-wheel drive Dutra D4K-A with equal-sized wheels appeared in 1961. The 65 hp D4K-A with a four-cylinder Csepel engine was advertised as a 'tractor with a power to weight ratio designed to give a high drawbar pull with almost complete elimination of wheelslip'. The Dutra D4K-B with a Csepel 100 hp six-cylinder diesel engine, ten forward and two reverse gears, four optional creep speeds and a top speed of 20 mph was launched in 1964. The advanced design of the D4K-B included a 540/1,000 rpm power take-off and air brakes on all four wheels. Having investigated the potential sale of Dutra tractors to western European farmers Technoimpex in Hungary introduced a 102 hp Perkins diesel-engined

version of the DK4-B to UK farmers in 1965. Marketed in the UK by the Guest General Trading Company at Brandon in Suffolk, the DK4-B cost £2,700, had ten forward and two reverse gears, diff-lock and category II hydraulic linkage. The tractor with a Csepel engine was available in the UK from 1966 but most buyers chose the Perkins-powered model.

Dutra Tractors (GB) Ltd, also at Brandon, imported four models of Dutra tractor in the early 1970s. The 6-354 and 6-354 Turbo had six-cylinder Perkins direct injection engines rated at 102 and 125 hp respectively, the 8-613 had an indirect injection 110 hp Csepel power unit and the Perkins-engined Dutra 6-372 was rated at 115 hp. The 6-354 and 8-613 tractors, also badged as the Dutra Landlord had a ten forward and two reverse gearbox, diff-locks on both axles, hydrostatic steering and an independent closed circuit hydraulic system with lower link sensing. The ten forward gears were split to give four creep speeds, four field speeds up to 7½ mph and two gears for roadwork. The clutch and diff-locks were pneumatically operated with compressed air rams connected to a small engine-mounted compressor. An optional assistor ram was available for the hydraulic linkage and releasing a simple locking mechanism on the left-hand lift rod made it easier to attach mounted implements. Sales literature explained that the tractors would 'perform effortlessly day in, day out pulling the heaviest equipment in the most difficult conditions'. The last Dutra tractors were made at Budapest in 1973.

Chapter 3

Ebro – Gunsmith

Ebro

Ebro tractors, named after a Spanish river, were made by the state-owned Motor Iberica SA at Barcelona. The company, which was formed in association with Ford in 1920, bought the production rights to the New Fordson Major from the Ford Motor Company in 1954 and made the tractors in Spain. A few Ebro tractors were sold to British farmers before it was replaced in 1958 by the Super Ebro, which was a Fordson Super Major in disguise.

97. Launched at the 1977 Royal Show the Ebro 460 with a Perkins AD4-203 engine and an eight forward and two reverse gearbox was one of ten 18–115 hp Ebro models that Motor Iberica planned to launch in 1978. (Stuart Gibbard)

When the Super Major was discontinued in 1965 Motor Iberica made the Super Ebro 55 under licence until 1970.

Following an agreement with Massey Ferguson in 1966 Motor Iberica also made the Ebro 155D, a copy of the Massey Ferguson 135 complete with a red MF bonnet and flat top wings, alongside the Fordson-based Super Ebro 55. The later Ebro 160D, a Motor Iberica version of the MF135, had a Perkins diesel engine and six-speed transmission. Massey Ferguson established their own assembly line in Barcelona linked by computer to the Banner Lane factory to build MF tractors for the Spanish market.

Motor Iberica used its UK importer Ferranti Engineering to test the British market. At the 1980 Royal Show they introduced three new Ebro 6000 models with an Ebro-designed twelve forward and four reverse synchromesh gearbox, hydraulic linkage and a 540/1,000 rpm power take-off. The two-wheel drive Ebro 6080, 6100 and 6125 had 78, 102 and 130 hp Perkins engines which were also made under licence by Motor Iberica in Spain. The four-wheel drive 6100DT and 6125DT, which cost about £22,500, had Italian-built Carraro front-wheel drive conversion kits and a de luxe cab was the only optional extra available for the Ebro 6000 series tractors.

Nissan bought the Massey Ferguson holding in Motor Iberica in 1979 and the earlier connection with Massey Ferguson resulted in the production of three Ebro models with 68, 76 and 90 hp Perkins engines at Barcelona in the early 1980s.

Eicher

Albert and Joseph Eicher used a 25 hp twin-cylinder Deutz engine for the Eicher T22 tractor which they made in 1936. During World War Two Eicher joined a consortium of German farm equipment manufacturers including Fahr and Güldner to build gas engines. They were making tractors with their own 17–19 hp single-cylinder air-cooled diesel engines by the late 1940s and the Eicher range was then extended to include two-, three- and four-cylinder engined tractors with a separate cooling fan for each cylinder. The Eicher Kombi tool carrier with a rear-mounted 15 hp Deutz diesel engine and a five forward and one reverse gearbox was launched in 1952. Later Eicher Kombi tool carriers had 11 and 19 hp Eicher air-cooled diesel engines.

The Eicher 19/20 with a single-cylinder water-cooled Deutz engine, five forward gears and one reverse and a 580 rpm power take-off was the smallest of the five Eicher models made in the 1950s. Air-cooled Deutz 25 and 30 hp twin-cylinder, 45 hp three-cylinder and 60 hp four-cylinder power units were used for mid-1950s Eicher tractors with a five forward and one reverse gearbox. Deutz used the same air-cooled engines for their own tractors.

The 13 hp Muli and the New Kombi tool carriers with 19 or 22 hp Eicher air-cooled engines appeared in 1957. The New Kombi had a five forward and one reverse gearbox but the Muli with six forward and two reverse gears had a 540 rpm rear power take-off and an unusual 982 rpm front power shaft with an optional reverse drive.

A twin-cylinder 26 hp four-wheel drive Eicher with equal-sized wheels launched in 1959 was one of ten models in an early 1960s range of tractors with air-cooled diesel engines in the 15–60 hp bracket. Some were named after big cats, such as the Leopard with a single-cylinder 15 hp engine and the twin-cylinder 22 and 29 hp Panther and Tiger. The larger 50 hp Mammoth I had three cylinders and the 55 hp Mammoth II had a four-cylinder air-cooled engine. The smaller tractors had six forward speeds and one in reverse, others had eight forward gears while the Mammoth I and II had four reverse gears and the Leopard had two.

A few Eicher tractors were sold in the UK, including the mid-1960s Puma with a 28 hp engine, six forward gears and category I hydraulic linkage, mainly to fruit farmers in the southeast. Narrow versions of the Puma with two- and four-wheel drive and two- or three-cylinder 30 and 50 hp engines were sold in the UK in the early 1970s along with improved versions of

98. The early 1950s Eicher 19/20 had leaf spring suspension on the front axle.

the Mammoth and the 52 hp Allrad four-wheel drive tool carrier. The two- and four-wheel drive Eicher Mammoth tractors had air-cooled turbocharged engines with a separate cooling fan for each cylinder, a dual clutch, twelve forward and four reverse gears and draft control hydraulics.

Massey Ferguson bought a share in Eicher in 1970 when some tractors were made with water-cooled Perkins engines. A turbocharged six-cylinder Eicher air-cooled power unit was used for the 95 hp four-wheel drive Wotan launched in the late 1960s.

99. The mid-1950s Fahr D17 had a direct drive from the engine for a mid-mounted mower.

Features of the Wotan included a sixteen forward and seven reverse gearbox, dual category lower link sensing hydraulics and front-wheel drive, which could be engaged or disengaged on the move. The more powerful Wotan 3135 with a 135 hp air-cooled Eicher engine had sixteen forward and seven reverse gears and drive to the front axle could be engaged or disengaged while on the move. The two-door cab was suspended on anti-vibration mountings.

However, the Massey Ferguson connection was short lived and Eicher's Indian division re-purchased the MF shares in the early 1980s. The Indian company thrived and in the late 1990s achieved an annual production of 20,000 German-designed two-wheel drive tractors with engines up to 40 hp. Eicher also made Trantor fast tractors in India in the mid-1980s and the new Trantor 904 was launched in 1999, mainly for the Indian and Chinese markets.

Fahr

Johan Georg Fahr began making farm machinery in 1870. Binders were added to the product range in 1910 followed by Fahr combine harvesters in the 1930s. The pale green twin-cylinder diesel-engined Fahr T20 with air-cooled Deutz or water-cooled Güldner engines made in 1939 was one of the first Fahr tractors. During the war years some had a gas-powered Deutz engine used with a wood gas generator.

Fahr Maschinfabrik introduced the first lightweight Fahr D series tractors in 1949. The D15, D17N and high clearance D17H had Güldner 15–17 hp twin-cylinder, water-cooled diesel engines. Other D series tractors included the 22 hp D22, which with the later addition of three extra horsepower under the bonnet, became the D25 and like the D28 and D30 they had a five forward and one reverse speed gearbox.

A mix of air- and water-cooled engines was used in the mid-1950s for Fahr wheeled tractors and tool carriers. The D270, D400 and D540 tractors had 32, 45 and 60 hp Deutz engines respectively and Güldner made the twin-cylinder 17 or 24 hp engine for the GT130 self-propelled tool carrier. The GT130's driving seat was at the rear and drive from the front-mounted engine was taken through a five-speed gearbox to the rear wheels.

The late 1950s saw the launch of red-painted Fahr D88, D131 and D133 tractors with 15, 17 and 25 hp Güldner engines respectively. The Fahr D177S with a four-cylinder 34 hp Mercedes-Benz diesel engine and an eight forward and four reverse speed gearbox was also made at the time. Fahr and Güldner joined forces in 1961. They formed a working relationship with Deutz and K-H-D, which already owned Deutz, became part owners of Fahr in 1968 and completed the takeover in 1970. Klöckner-Humboldt-Deutz used the Deutz-Fahr name on their tractors for the first time in 1981 when they launched the four-wheel drive Deutz-Fahr DX and 07C medium-powered tractors.

Farm Tractor Drives

Arthur Battelle formed Farm Tractor Drives Ltd at Shardlow in Derbyshire in 1977 in order to import Schindler front-wheel drive kits and transmission units from Switzerland. More than 500 FTD four-wheel drive kits

100. The FTD Chieftain was designed for farmers wishing to buy an economically priced four-wheel drive tractor.

for Ford and Massey Ferguson tractors had been sold when the new four-wheel drive FTD Chieftain appeared at the 1981 Royal Smithfield Show.

Limited production of the low-cost, no-frills Chieftain made with new and re-manufactured Ford tractor components, a six-cylinder 105 hp Ford 2725 Economy industrial engine and a rebuilt Ford 5000 or 7000 transmission with eight forward and two reverse gears started in 1982. The Schindler front axle with a torque-biasing differential enabled the front wheel with the most traction to absorb up to three times as much torque as the earlier Schindler front axle. The Chieftain complete with power-assisted steering and a Duncan safety cab with a radio and heater cost £13,750 which was about £5,000 cheaper than many two-wheel drive tractors on the market at the time. Optional extras included a front weight frame, an assistor ram for the hydraulic linkage, Dual Power, Load Monitor and an auxiliary 18-gallon fuel tank. Second-hand components and a Schindler front-wheel drive conversion kit were used for the cheaper four-wheel drive FTD Warrior.

The major tractor manufacturers showed very little interest in four-wheel drive at the time and production of conversion kits was left to County, Roadless and a few other manufacturers making 100 hp-plus tractors for large arable farms. This left an opening for FTD to

sell the keenly priced Chieftain to farmers with small to medium-sized areas of grassland or mixed arable crops. However, David Brown initiated a price war in the early 1980s and within a couple of years the Chieftain had lost its price advantage and with only a handful built the last one was sold in 1985.

Farmwell

Workwell Engineering at Ashford in Kent made two prototype Farmwell 467 tractors, based on 60 hp Massey Ferguson 265 skid units in 1985. Designed for under-developed countries the Farmwell was a very basic tractor on steel wheels without any electrical equipment and started with a wind-up inertia starter instead of the normal electric starter motor. Inertia starters, made by CAV and Simms Motor Units, had a crank handle used to compress a heavy spring and when it was fully wound a locking mechanism was released and the expanding spring turned the starter pinion and flywheel ring gear in the normal way.

An 80 hp Massey Ferguson 290 skid unit was used to build the similar steel-wheeled Farmwell 480 and about 300 of these tractors were shipped to Africa in knockdown form in the late 1980s and early 1990s. The Farmwell was not a pretty tractor but the 3 mm thick sheet steel bonnet and body panels protected the fuel tank, radiator and other vulnerable engine

101. The Farmwell 480 was based on a Massey Ferguson skid unit.

components. An hour meter was its only instrument and a single grade of oil was specified for use throughout the tractor. A limited number of the pneumatic-tyred Farmwell 6110 tractors based on Massey Ferguson 399 skid units with 100 hp Perkins engines were sent to Africa for sugar cane production and other heavy haulage work.

Workwell Engineering also made Tugwell and Loadwell 480 tractors at Ashford. Similar to the Farmwell but without the hydraulic linkage the Tugwell had an inertia starter, pneumatic tyres and a heavy-duty drawbar. The Loadwell was based on the Tugwell with a rear ballast weight, an auxiliary hydraulic system and a heavy-duty front-end loader.

Fendt

Johann-Georg Fendt and his son Hermann made a four-wheel self-propelled grass mower with a 4 pk (hp) petrol engine in 1928 to cut the farm hay crop. As it lacked power they built the 6 pk (hp) steel-wheeled Dieselross (Diesel Horse) tractor in 1930 and used it with various implements including a side-mounted mower and a mounted plough. Pk or ps is the German abbreviation for pferdestarken and like the French

chevaux-vapeur (cv) is one metric horsepower which is equivalent to 1.014 imperial hp.

The F9 Dieselross with a 9 hp horizontal single-cylinder crosswise-mounted diesel engine, a sprung front axle and solid tyres appeared in 1932. The brothers Hermann, Xaver and Paul Fendt went into partnership as Xaver Fendt & Co at Marktoberdorf in 1937 when they launched the single-cylinder 16 hp Fendt Dieselross F18 with an independent power take-off which could be disengaged under load. The Dieselross F22 with a vertical twin-cylinder water-cooled 22 hp engine and four-speed gearbox which appeared in 1938 was the last Fendt diesel tractor introduced before the onset of war. A shortage of fuel in Germany resulted in a ban on the use of diesel fuel for tractors so Fendt made a wood gas generator for the Dieselross tractor.

Many German tractor and farm equipment manufacturers, including Xaver Fendt & Co, spent the immediate postwar years rebuilding their industry. The Fendt 18 hp Dieselross F18H with six forward speeds and one reverse launched in 1949 was the last Fendt tractor with a single-cylinder horizontal engine made at Marktoberdorf. A new era of Fendt tractors started in

102. Fendt launched the F18 Dieselross in 1949.

1950 with the introduction of the F15 Dieselross with a single-cylinder vertical 15 hp MWM engine. The low-profile F15's short wheelbase gave it an old-fashioned look but the tractor's optional eight forward speed gearbox was well ahead of its time.

A new generation of Dieselross tractors with 12–40 hp engines appeared in 1952 and the 28 hp Fendt F28 with a twin-cylinder water-cooled MWM diesel engine and a 12-volt electric starter was typical of the range. However, a starting handle was provided for emergency use! The tractors had a five forward and one reverse gearbox with optional high speed and creep speed ratios, independent servo-assisted rear drum brakes, a hand brake for the transmission, diff-lock and power take-off. The Fendt transverse leaf spring suspension on the front axle used on earlier models was retained and the steering was said to be 'light enough to control with one hand'. There was a belt pulley on the side of the transmission housing behind the rear axle and a direct drive from the engine was provided for a side-mounted mower, which many farmers kept permanently attached to the tractor. Optional equipment included hydraulic lift and

linkage, a weather cab, flashing indicators, a pto-driven tyre pump and an electric hand lamp.

The first Fendt GT tool carrier, launched in 1953, had a 12 hp MWM diesel engine mounted above the rear axle and four implement attachment points. The ageing Dieselross tractors were replaced with the twin-cylinder Farmer and three-cylinder 40 hp Favorit I in 1958. Fendt celebrated the production of their 100,000th tractor in 1961 and improved versions of the Farmer range appeared throughout the 1960s. A new self-propelled toolbar with front-, mid- and rear-mounted implement attachment points and a 45 hp three-cylinder diesel engine was launched in 1965. The Farmer 3S Turbomatik range, introduced in 1968, set a new trend for Fendt tractors. The Turbomatik turbo-clutch or fluid flywheel was used to transmit drive through a single-plate clutch to the gearbox with the engine set at a pre-selected speed.

Bill Bennett Engineering at Chipping Sodbury near Bristol introduced Fendt tractors to British farmers in 1974. Early imports included the two-wheel drive 105 hp Favorit 610S and the two- or four-wheel drive six-cylinder 120 hp 611S and 612S which had water-cooled direct injection engines with individual cylinders for easy servicing. A sales leaflet explained that the Fendt 612S was 'built to tackle big acreages and the heavy jobs year after year'. The 610S had twelve forward and five reverse gears and a sixteen forward and seven reverse speed transmission was standard on the 611S and 612S. Standard features included the Fendt turbo-clutch, power steering, a sprung front axle and an external assistor ram for the hydraulic linkage, which could be used in position, draft control or a mix of both.

The two- and four-wheel drive, three-cylinder 55 hp Farmer 104S and the four-cylinder 65 hp Farmer 106S, also available in the UK in 1974, had a synchromesh gearbox with an optional reverser unit for quick directional changes when using a front-end loader. The 36 and 50 hp rear-engined 250GT (Gerate Trager or tool carriers) with a safety cab completed the Fendt tractor range in the UK. Tractors for vineyards, hop cultivation and orchard work were added in 1975.

103. The 12 hp Fendt Dieselross was introduced to German farmers in 1952.

British farmers had an even wider choice of Fendt tractors in 1976 including the 85–150 hp Favorit models, the Farmer range and two GT tool carriers with 55 and 78 hp engines. Five new Fendt Favorit 600S and 600SL Turbomatik models introduced later that year had 100–165 hp direct injection engines. Other

104. Bennett Engineering were importing the Fendt 250GT tool carrier in 1974.

105. The Favorit 626LS Turbomatik had 252 hp under the bonnet.

features included independent power take-off, lower link sensing hydraulics with automatic couplers and a heated quiet cab which could be tilted backwards in order to service the transmission and hydraulic system. A Bill Bennett Engineering advertisement claimed that 'more and more farmers are finding that high-quality Fendt tractors give them a new standard of reliability and operator satisfaction'. The 55 hp three-cylinder Fendt 255GT tool carrier and the four-cylinder 70 hp GT275 with Deutz diesel engines were current in 1984. Both models, with a fourteen forward and four reverse transmission, could be used with front-, mid- and rear-mounted implements including a crop sprayer with the tank on the front load platform and a rear-mounted spray bar.

A twelve forward and five reverse gearbox was used on the 100 hp Favorit 610S, while the other Favorits had a sixteen forward and seven reverse speed gearbox. An optional creeper box provided an additional eight forward and four reverse speeds. The more powerful 211 hp

622LS and the 252 hp 626LS four-wheel drive tractors were added to the Favorit range in 1979. The flagship 626LS had eighteen forward and six reverse speeds provided by six forward and two reverse synchromesh gears and a clutchless three ratio change in each forward and reverse gear. A sideways tilt adjustment for the cab made it possible for the driving seat to remain level when ploughing with one wheel running in a furrow. Other features included electro-hydraulic control of the hydraulic linkage and diff-locks.

106. The Farmer 310 had a twenty-one forward and six reverse speed transmission.

The restyled two- and four-wheel drive Farmer 300 Turbomatik range launched in 1980 had four-cylinder 50–86 hp engines, an improved fourteen forward and four reverse synchromesh gearbox, four-wheel braking and an optional 40 kph (25 mph) transmission. The safety cab on rubber mounting blocks, a two-speed power take-off and an automatically engaged diff-lock were included in the price of the 78 hp 308LS and 86 hp 309LS. The 92 hp turbocharged four-cylinder Farmer 310 with twenty-one forward and six reverse gears was added in 1984. The Fendt Farmer's improved Fentronic hydraulics with electronic linkage control were operated with a joystick in the 'climate-controlled workplace'. The six-cylinder 115 hp 312LSA introduced in 1987 was similar to the other Farmer models and the two- and four-wheel drive 250S, 260S and 275S launched in 1988 were the first of the new Fendt 200 series. The 312LSA gearbox was used for the new Farmer 50, 60 and 75 hp tractors and the 40 hp 240S was added within a year. Fendt introduced a factory-fitted reverse drive for the Favorit range at the 1989 Royal Show. It was suitable for Favorit models with either a synchromesh or Duo-speed transmission and it was only necessary to swivel the driving seat to switch the controls to the rear of the tractor automatically.

Fendt enjoyed a 20 per cent share of the West German tractor market when a new range of the four-wheel drive Fendt GTA tool carriers with 65, 80, 100 and 115 hp air-cooled engines appeared in 1991. The specification included the Fendt electronic hydraulic control unit with a shock load stabiliser system to reduce implement bounce on the rear linkage and two-wheel drive option was available on the 65–100 hp GTA models.

There were twenty Farmer 200 and 300 series tractors, four 600 series Favorit models and eight Fendt tool carriers in production in 1992 when Bonhill Engineering Ltd of South Cave in Yorkshire were appointed Fendt distributors for the UK. Restyled and more powerful Farmer 300LS and LSA tractors with sloping bonnets and a new front grille were launched in the same year. The smallest Farmer 304LS/LSA had a three-cylinder 70 hp engine while the six-cylinder 312LSA rated at 120 hp was the most powerful model in the range. An optional twenty-one forward and reverse speed shuttle gearbox with an electro-magnetic pre-selection system was available for Farmer tractors over 75 hp.

The new 105–140 hp 500 series and 165–230 hp 800 series Favorit tractors announced in 1993 had the Fendt Turboshift single lever shuttle transmission system with twenty-four forward and reverse work speeds with an optional twenty creeper gears in both directions. The 800 series had six-cylinder MAN engines, hydro-pneumatic front axle suspension with an optional levelling control, electronic front and rear hydraulic linkage control and a sprung suspension cab. The Farmer 300 series was revised again in 1995 when the six models had three-, four- or six-cylinder engines in the 75–125 hp bracket. Electronic control systems similar to those on the Favorit range were available at extra cost and the 75 hp Farmer 300 was the only four-wheel drive model without optional front axle suspension.

Fendt exhibited the Xylon 320 systems tractor at the 1990 Royal Smithfield Show to test the reaction of their farmer customers. The Xylon with the engine and cab mounted amidships had front and rear three-point hydraulic linkages, a load-carrying platform and provision for mid-mounted implements. The specification included a 120 hp six-cylinder air-cooled engine, a twenty-one forward and six reverse gearbox and a tiltable cab with two seats. However, advertisements advised potential buyers to keep their chequebooks in their pockets, as it would be at least two years before the Xylon would be available in the UK. The first production models of the Xylon tool carrier or 'systems tractor' appeared at the 1994 Royal Smithfield Show when there was a choice of colour schemes for the Xylon 520, 522 and 524 with 110, 125 and 140 hp MAN power units. The transmission system included the Turbomatik clutch and a four-stage powershift gearbox with pre-select reverse giving twenty-four field and twenty creep speeds from 0.5 to 50 kph. Other features included a swivel-joint front axle with lockable suspension and a hydraulic tilt facility for the two-seater cab.

When AGCO, which owned Massey Ferguson, acquired Xaver Fendt & Co in 1997 they took over the distribution of Fendt tractors in the UK and within a year a number of MF dealers were also selling the German-built tractors. Having lost the Fendt franchise, Bonhill Marketing was appointed the UK distributor for Steyr tractors in 1998.

Fendt introductions in 1997 included the 900 Vario series, the 150 hp Favorit 515C, an improved

107. There was a choice of colour schemes for the mid-1990s Fendt Xylon.

transmission for the 900 series and the Farmer 307C, 308C and 309C rated at 75, 85 and 95 hp respectively replaced equivalent Farmer 300LSA models. The combination of a mechanical gearbox and a new stepless hydrostatic transmission with a single joystick control was a special feature for the 170, 200, 230 and 260 hp 900 series Vario tractors with MAN six-cylinder engines. Fendt introduced the six-cylinder 140 and 160 hp Favorit 714 and 716 with Deutz engines at the 1998 Royal Smithfield Show. They were the first of a new range of Vario Favorit 700 series tractors with a continuously variable transmission (CVT), front axle suspension and a hydro-pneumatic suspended cab.

There were twenty-five different tractors in the 2001 Fendt price list. They included the Farmer 300, 400 and 500 Series Varios, Favorit 700 and 800 Vario tractors, the 370GT and 380GTA tool carriers, two Xylon systems tractors and a range of vineyard and fruit models. The smallest 250V two-wheel drive vineyard tractor with a 50 hp engine and a

twenty forward and six reverse gearbox cost £22,600 and the flagship four-wheel drive Fendt Favorit 926 Vario with 260 hp six-cylinder engine and CVT had a £95,000 price tag. An advertisement claimed that 'the 926 Vario offers permanent peak performance and the Variotronic monitoring system reduces operator stress to keep the driver fresh and alert'.

108. The six-cylinder Fendt Vario 700 range with a stepless transmission was launched at the 1999 Agritechnica Show.

Three new 800 TMS (Tractor Management Systems) Vario tractors with six-cylinder 168–195 hp engines and CVT and a new 310 hp flagship Vario 930 were announced in 2002. The joystick-controlled TMS, which included a headland management system, was designed to ensure the engine and Vario transmission remained in the most efficient operating range. The 2003 Fendt range included the Farmer 200 and 300 tractors with three- or four-cylinder 65–112 hp power units and a twenty-one forward and six reverse gearbox. The Farmer 400 and the Favorit 700, 800 and 900 Vario models had a 92–318 hp engines and CVT transmissions. The 800 Vario TMS series had twenty-four valve electronically managed engines and one touch on the joystick started a sequence of up to thirteen driving functions. A reverse drive option was available for the flagship 310 hp 930 Vario TMS launched at the 2003 Royal Highland Show. Features included the Variotronic TI headland management and TMS. The Variotronic headland system stored up to sixteen headland turning sequences and up to thirteen functions for four different implements were repeated with the touch of a button and the TMS linking the engine to the transmission ensured that it ran at the most efficient speed.

Ferguson

Harry Ferguson, christened Henry George, was born in County Down in 1884. His first experience with tractors was in Ireland in the 1920s from where he sold the American Waterloo Boy Model N in Northern Ireland as the Overtime Model N. The first Ferguson with an American Hercules four-cylinder side-valve petrol engine was made in 1933, which because of its colour, was known as the Ferguson Black tractor. An agreement with David Brown resulted in the manufacture of the Type A or Ferguson-Brown tractor at the Park Works in Huddersfield between April 1936 and June 1939. A 20 hp Coventry Climax side-valve engine, similar to the Hercules engine used for the Black tractor was used for the first 200 or so battleship grey Ferguson Brown tractors with the patented Ferguson hydraulic weight transfer and draft control system.

From 1937 the engine for the Type A, which was similar to the earlier Coventry Climax power unit, was made by David Brown at Huddersfield. Farmers were offered the choice of standard 10 in or narrow 6 in wide rowcrop steel wheels or pneumatic tyres in 1937 and an optional vaporiser kit was introduced in 1938. Four implements, including a plough and a cultivator, were made for the

109. The Ferguson Type A was made between 1936 and 1939 by David Brown at Huddersfield in Yorkshire.

110. A number of Ford Ferguson 9N tractors were sent from America to Britain during World War II.

Ferguson Brown tractor and Harry Ferguson insisted that only two sizes of high-tensile bolts with hardened heads were to be used so that the Ferguson spanner would fit all nuts and bolts on the tractor.

Anticipating the break-up with David Brown, Harry Ferguson concluded the famous 'gentlemen's agreement' with Henry Ford to build Ferguson tractors in Detroit. About 1,350 Ferguson Brown Type A tractors had been made when it was discontinued in 1939 and Henry Ford made the first Ford Ferguson 9N tractors in America in the same year. The Ford 8N introduced in 1947 was virtually identical to the Ford Ferguson 9N and the resulting lawsuit dragged on for four years.

Ferguson, faced with the loss of his American market after the break up with Ford, bought a 72-acre factory site in Detroit and the first TO20 (Tractor Overseas) tractors were made at Ferguson Park on 11 October 1948. The TO20 which was similar to the TE20 was superseded by the 30 hp TO30 in 1951. The TO35 which replaced the TO30 in 1954 was made for six years.

The Ford Motor Co made about 80 per cent of the tractors produced in the UK between 1936 and 1946 but the launch of the TE20 (Tractor England) heralded a revolution in farm mechanisation. The first Ferguson tractors made at the Standard Motor Company's Banner

Lane factory in Coventry in July 1946 were similar to the 9N with a 24 hp overhead-valve petrol engine manufactured by the Continental Motor Co at Michigan in America. The TE (Con) 20, as it was known, had a four forward and one reverse gearbox with a safety-start mechanism built into the gear lever, reshaped rear wheel centres, a front-hinged bonnet and a 1⅛ in diameter power take-off shaft. The tractor also had a master brake pedal acting on both wheels and independent steering brake pedals. The stock of Continental engines ran out in 1947 and with a new power unit based on the Standard Vanguard car engine the tractor became the Ferguson TEA20 with a 24 hp petrol engine.

Double shift work began at Banner Lane in 1948 and production was increased to about 5,500 tractors per month to meet a huge demand for Ferguson tractors. Launched in 1949 the TED20 was, except for its vaporising oil engine, identical to the TEA20. The TEH and TEJ lamp oil models were added in 1950 and engine modifications in 1951 uprated the TEA20 to 28 hp and the TED20 had 26 hp under its bonnet. The original 6-volt electrical system was changed to 12 volts and the four-cylinder 26 hp TEF20 diesel tractor with a heavy-duty 12 volt starter and a safety switch on the side of the gearbox was launched the same year. A decompression

111. More than half a million standard, narrow width and industrial Ferguson TE20 tractors were made at Banner Lane in Coventry

lever and a Ki-gass system were provided to help start the TEF20's indirect-injection engine in cold weather. A three-cylinder Perkins diesel engine conversion was introduced in 1952 for the petrol- and paraffin-engined tractors. The Ferguson catalogue for that year included fourteen models of the TE20 tractor, thirty-four Ferguson implements and a range of twenty different accessories including a tractor jack and tyre inflation kit.

Various gadgets were made during this period to make the Ferguson 20 easier to use. Spanners were needed to alter the length of the standard Ferguson top link but the screw-adjusted Speedy top link made by Tamkin Bros of Maldon in Essex with a circular handle made this adjustment a simple task. A large boot was useful when the driver needed to use the clutch and the left-hand steering brake pedals at the same time but the Tamkin Slewstick made life a little easier for the driver. It consisted of a tubular handle attached to the left-hand steering brake pedal which could be used to apply the left-hand brake while the left foot remained in full

control of the clutch pedal. The Gray-Thompson safety clutch-release made by Grays of Fetterangus in Scotland was designed to stop the tractor if a mounted implement hit a buried object. The device, which cost £7.10s, consisted of a spring-loaded mechanism on the top link connected by a mechanical linkage to the clutch pedal. When an implement hit an obstruction the shock load was transmitted through the top link which released the overload mechanism and disengaged the clutch.

A full range of Ferguson implements was available for the tractor but as Harry Ferguson lacked the necessary manufacturing facilities production was sub-contracted to several other companies. They included Rubery Owen at Darlaston who made the ploughs and cultivators, Midland Industries at Wolverhampton who produced the potato planter, Scottish Mechanical Light Industries who made the Ferguson hammer mill and James Sankey & Sons at Wellington who manufactured tipping trailers.

Ferguson TE Tractor Models 1946–56

Model	Build	Fuel	Model	Build	Fuel	Model	Build	Fuel
TE20 (Con)	Standard	Petrol	TEF	Standard	Diesel	TEP	Industrial	Petrol
TEA	Standard	Petrol	TEH	Standard	Lamp Oil	TER	Industrial	Vaporising Oil
TEB20 (Con)	Narrow	Petrol	TEJ	Narrow	Lamp Oil	TES	Industrial	Lamp Oil
TEC	Narrow	Petrol	TEK	Vineyard	Petrol	TET	Industrial	Diesel
TED	Standard	Vaporising Oil	TEL	Vineyard	Vaporising Oil			
TEE	Narrow	Vaporising Oil	TEM	Vineyard	Lamp Oil			

Massey-Harris and Ferguson joined forces in 1953 and Massey-Harris-Ferguson continued production of TE20 tractors until 1956 by when more than 517,000 grey Fergusons had been made at Banner Lane. Ferguson tractors achieved average annual sales in excess of 20,000 in Britain and many more were exported to Scandinavia, Australia and other major customers.

The grey and gold Ferguson FE35 with a four-cylinder Standard engine which depending on build could run on petrol, vaporising oil, diesel or lamp oil and was rated at 37, 30, 37 and 29 hp respectively was launched at the 1956 Royal Smithfield Show. The FE35 had a safety-start mechanism built into the range gear lever, a six forward and two reverse gearbox, engine and ground speed power take-off and a new two-lever hydraulic system with position control, response control and two-way draft control. The de-luxe FE35 had a dual clutch with live hydraulics and power take-off. The new tractor also had the same hinged bonnet with a separate access panel for daily engine maintenance as the earlier TO20. A tip-back cushioned-seat with a backrest improved driver comfort and additional instruments recorded engine hours, registered power take-off speed and measured forward travel in each gear. An American publicity film praised the tip-back seat which it was suggested made it 'possible for the operator to drive while standing on the foot plates', presumably to take a break from sitting on the not quite so comfortable seat cushion.

Harry Ferguson relinquished his interest in M-H-F in 1957 and Harris was dropped from the company name. The grey and gold FE35 became the Massey Ferguson 35 with a new red and grey colour scheme and the MF triple triangle badge. The same badge and colour scheme was used for the MF65 launched at the 1957 Royal Smithfield Show.

112. Although the grey and gold Ferguson FE35 with a four-cylinder diesel engine had a pre-heater in the inlet manifold, it was not the easiest tractor to start on cold days.

Ferrari

Not connected to the Ferrari Car Company Fernando Ferrari was making pedestrian-controlled garden cultivators in the mid-1950s. Bogie seats were added in the early 1960s followed by pivot-steered and conventional ride-on front-wheel steered tractors.

The Ferrari 75RS and 30W/S DT were current in the mid-1970s. The equal-sized four-wheel drive 7RS had an air-cooled twin-cylinder 30 hp diesel engine, a seven forward and three reverse speed gearbox and diff-lock. The 30W/S DT was a conventional two- and four-wheel drive tractor with a water-cooled diesel engine and a twelve forward and six reverse gearbox.

A range of six Ferraris including the four-wheel drive Ferrari 95, 85 and 86 articulated tractors with safety frames were available in the UK in the early 1980s. The Ferrari 95, with the option of a 33 hp twin-cylinder or 34 hp three-cylinder engine, had a six forward and three reverse gearbox, engine and ground speed power take-off and hydraulic linkage. The three-cylinder Ferrari 85 and 86 with 45 and 55 hp engines respectively, had the option of single or dual clutch, a seven forward and three reverse gearbox, power-assisted steering and hydraulic linkage described in sales literature as 'an independent hydraulic implement lifter with depth control'.

Ferrari UK at Oldham in Lancashire sold pedestrian-controlled garden tractors, ride-on lawnmowers and four-wheel drive tractors in the early 1990s. The largest and most expensive pivot-steered tractors with 49 and 64 hp three-cylinder diesel engines had a sixteen-speed synchromesh gearbox and oil-immersed brakes. These reverse-drive Ferrari tractors could be used in either direction as the driving seat, steering wheel and controls could be rotated through 180 degrees. They could be used with rear-mounted implements or when driven in reverse with a mower on the rear linkage it became an articulated front mower. The more conventional four-wheel drive 29 and 38 hp Ferrari Model 22 and 40 diesel tractors had twelve forward and six reverse gears, two-speed power take-off and hydraulic linkage.

Fiat

The Italian Fiat company, founded by Giovanni Agnelli in Turin in 1899 to manufacture motor cars, took its name from the initial letters of Fabbrica Italiana di Automobili Torino. The first Fiat 702 tractors made in 1919 had a four-cylinder water-cooled Fiat engine, which developed 30 hp when running on petrol, but this dropped to 25 hp on paraffin. Features included a three forward and one

113. The articulated Ferrari 74 had an air-cooled diesel engine.

114. Introduced in 1919 the 25 hp Fiat 702 had a sprung front axle and an offset driving position to improve the forward view when ploughing.

115. *Electric starting was standard on the diesel-engined 70 hp Fiat 70C crawler; a donkey engine was available as an optional extra.*

reverse gearbox, a three-speed belt pulley, a sprung front axle and an offset driving seat which gave a better forward view when ploughing. Fiat Tractors at Abermarle Street in London, who marketed the Italian tractors in the UK between 1919 and 1925, demonstrated the steel-wheeled 702 at the 1919 Lincolnshire Tractor Trials. The judges were impressed with the tractor's fuel economy and its ploughing rate of 1¼ acres an hour with a six-furrow plough set at an 8 in depth on light sand. The 70 hp Fiat 70C crawler introduced in 1932, was Europe's first mass-produced tracklayer. There was a shortage of good-quality fuel in Italy when Fiat introduced the 41½ hp model 41 Boghetto crawler in 1938 and to overcome the problem the tractor had a multi-fuel engine that could run on petrol, fuel oil, diesel, natural gas, low-octane gas or alcohol.

Tractor production resumed in 1946 when Fiat launched the Model 50 and 52 crawler tractors which were the first Fiat tractors with a full diesel engine. Started with a 10 hp two-stroke donkey petrol engine the 50 hp four-cylinder engines had two inlet and two exhaust valves for each cylinder. The water-cooling system, which it shared with the donkey engine, provided hot water to pre-heat the cylinders before starting the main engine. Both models had five forward gears and one reverse and a steering wheel to control the multi-plate steering clutches. The Model 52 crawler had slightly lower gear ratios and a 59 in track centre compared with the 46 in track centre on the Model 50.

Steel wheels were optional for the 22 hp petrol or 18 hp vaporising oil-engined Fiat 600 with a four forward speed and one reverse gearbox launched in 1949. Variants of the Fiat 600 included the 602 tricycle-wheeled rowcrop model with adjustable wheel track settings and the 601 tracklayer with standard tracks for farm work and narrow tracks for orchards and vineyards. Standard equipment included a starting handle and a 6-volt lighting system with a friction-wheel driven dynamo which could be disengaged when the lights were not required.

The Fiat 55 crawler announced in 1950 had a 55 hp diesel engine started with a 10 hp twin-cylinder horizontally opposed donkey petrol engine bolted to the flywheel housing and used in the same way as an electric starter motor with a lever-operated clutch mechanism. This engaged the donkey engine starter pinion with the flywheel ring gear and turned the diesel engine while on full compression. The Fiat 55 with a ground pressure of less than 5 psi cost about £2,600. It had five forward gears and one reverse with a top speed of 5 mph and, like the earlier Model 50 and 52, a steering wheel was used to control the clutch and brake steering system.

Fiat introduced sixteen new models of two- and four-wheel drive tractors and five crawlers during 1959 and 1960. The 200, 300 and 400 series with 22–85 hp diesel engines and the petrol-engined Fiat 211 wheeled tractors were not available in the UK. The Fiat 311C, 411C and 70C crawlers with 30, 40 and 70 hp diesel

116. *The six-cylinder Fiat 100C crawler cost £11,450 ex works in 1975; hydraulic linkage, power take-off and a belt pulley were optional extras.*

engines were imported by McKay Industrial Equipment Ltd at Feltham in Middlesex in the early 1960s. The more popular 70C, made between 1961 and 1971, had a four-cylinder overhead-valve engine with a 24-volt electric starter, five forward and four reverse gears with a quick reverse lever and multi-plate steering clutches. A toolkit and a grease bucket were included in the basic price but the power take-off, belt pulley and an alternative 10½ hp donkey starting engine were optional extras. Manns of Saxham were appointed Fiat crawler tractor distributors for England and Wales in 1963.

Fiat established a tractor and earth-moving equipment division in the late 1960s and opened an office at Berkeley Square in London in 1972 to market their crawler tractors in Britain. The 80 and 100 hp 80C and 100C tractors launched in 1971 had four-cylinder direct injection engines with a 24-volt electric starter. Other features included an oil-cooled dual-plate clutch, five forward gears and one reverse, clutch and brake steering and hydraulic track adjustment. The smaller 65 hp French-built Fiat 655C had a six forward and two reverse gearbox and a 12-volt electrical system.

The Fiat Trattori agricultural division was established in 1974 and the Fiat-Allis Construction Equipment Ltd formed a joint venture with Allis-Chalmers in the same year. Laverda joined Fiat Trattori in 1975 and Hesston became part of the group in 1977. Fiat-Allis Construction Machinery (GB) Ltd were marketing Fiat 80C, 850C and 100C crawlers and Fiat 540, 640, 850, 1000 and 1300 two- and four-wheel drive (DT) tractors

from the mid-1970s. The smaller models rated at 54, 64, and 85 hp had four-cylinder water-cooled direct injection engines and 100 and 130 hp six-cylinder engines were used for the Fiat 1000 and 1300. The 540 and 640 had eight forward and two reverse gears, the others had twelve forward and four reverse speeds in a three-range gearbox. Hydrostatic steering was standard on the Fiat 1300 while the other models had power-assisted steering. Nine models of the Fiat 80 series tractors in the 58–80 hp bracket appeared between 1973 and 1979 and a super comfort quiet cab styled by Pininfarina was added to the list of optional equipment in 1978.

The 150 hp Fiat 1300 Super and four-wheel drive 1300DT Super with flat floor quiet cabs replaced the 1300 and 1300DT in 1977. The 110 hp Fiat 1000 Super and 1000DT Super replaced the 1000 and 1000DT in the same year. With the extra horsepower provided by increased fuel injection pump capacity and higher engine speeds the new models were the most powerful tractors built in Italy at the time. They retained the lower link sensing hydraulics and telescopic lift arm ends used on earlier models. They were also the first Fiat tractors with an external remote hydraulic control switch for the driver to use when attaching an implement. The new Super models, with an optional four-speed creeper box, had the same epicyclic final drive reduction units and the side-mounted drive shaft from a transfer box to the front wheel box as the earlier Fiat 1000DT and 1300DT.

There were sixty-two models in eighteen power bands from 28 to 350 hp in the Fiat range in 1979 when the Turin factory celebrated sixty years of tractor production. They included the earlier 540, 640 and 850 models already re-launched as the 58–88 hp 80 series together with four new two- and four-wheel drive tractors ranging from the 115 hp Fiat 1180 to the 180 hp 1880 launched in the Turin factory's Diamond Jubilee year.

Demand for high horsepower tractors in Europe was growing in 1978 when about 9,800 tractors over 180 hp were sold in North America but only 1,700 in the rest of the world excluding Eastern Europe. The apparent

117. The 280 hp articulated Fiat 44-28 with twelve forward and four reverse gears was made by Versatile in Canada.

sales opportunities in Europe resulted in a trading agreement with the Versatile Tractor Co in 1979 to supply articulated four-wheel drive tractors in Fiat colours. The Fiat 44-23, 44-28, 44-33 and 44-35 with a twelve forward and four reverse gearbox and category III & IV hydraulic linkage made by Versatile were launched in time for the 1979 Royal Smithfield Show. The four articulated tractors had 230, 280, 330 and 350 hp Cummins 'Constant Power' six-cylinder turbocharged engines.

The 80 series was extended in 1981 when Fiat launched the 1880DT and the 88 hp five-cylinder 880-5 and the 880-5DT which replaced the Fiat 880 and 880DT. The 180 hp 1880DT with a six-cylinder turbocharged engine had a twelve forward and four reverse gearbox, category II hydraulics with lower link sensing and automatic Walterscheid couplers on the lower lift arms. The 1800DT specification included hydrostatic steering and brakes and a 1,000 rpm power take-off. The Fiat 1280 and 1280DT with 125 hp turbocharged engines under the bonnet and a hydraulic power take-off clutch were added in 1982.

The Fiat 665C and 1355C crawlers with 68 and 135 hp engines were announced at the 1983 Royal Show. The 665C was based on the four-cylinder, eight forward and four reverse speed Fiat 680 wheeled tractor. It had

clutch and brake steering, hydraulically tensioned tracks, independent power take-off and three-point linkage. An unusual gull-wing two-door cab was included in the basic price of £13,000. Optional equipment for the 665C included a sixteen forward and eight reverse creep speed transmission or an eight forward and reverse shuttle gearbox. The Fiat 1355C with a conventional air-conditioned two-door cab had the same six-cylinder engine as the 1580 wheeled tractor with five forward and five reverse gears and Turner hydraulic linkage. Provision was made to attach an optional one tonne front ballast weight. Fiat Trattori became Fiatagri in 1984 when the 55–180 hp 90 series tractors gradually replaced the ageing 80 series. There were eleven two- and four-wheel drive Fiat 90 series from the 55-90 to the 180-90, the first two or three figures in the model number denoting engine horsepower. A comfort cab was standard on the smaller Fiats but there was a choice of a super comfort (S), comfort wide (CW) or super comfort wide (SW) cab for the 115–180 hp tractors. A twelve forward and four reverse synchromesh transmission was standard but a sixteen speed shuttle box with optional powershift was available for the 130-90 and larger models.

The Fiat 100-90 was the odd one out with a fifteen forward and three reverse gearbox or alternatively a twenty forward and four reverse gear and optional creep speed box. The flagship Fiat 180-90 distinguished itself in 1984 when it set a world record by ploughing an acre in 11 minutes 21.8 seconds. The 180 hp Fiat 180/55 crawler version with hydrostatic transmission and joystick controls was introduced at the 1986 Royal Smithfield Show.

Fiatagri had launched the medium horsepower 65 and 66 series tractors in 1984 and 1985 but the three-cylinder 45 hp 45-66 and 45-66DT were the only models included in the price list when Fiatagri UK opened a retail depot at Bury St Edmunds in Suffolk in 1986. The twelve forward and three reverse speed Fiat 45–66 had multi-disc wet brakes and Fiat Lift-o-Matic hydraulics with a four-position sensing control system. A creep speed box and a twelve-speed shuttle gearbox were optional extras. The 1986 Fiat crawler range included the four-cylinder 70 hp 70-65, the six-cylinder 1355C and the new Turbo-Hydro drive 180-55 with an automatic transmission with an over-ride pedal which could be used to vary forward speed to suit the load or gradient. The tracks were driven with hydraulic motors and a joystick was used to select gear ratios and steer the tractor.

There were more than thirty different makes and approximately 700 models of farm tractor on the British market including at least seventy with a Fiat or Ford

118. Fiat 1380 DT had a 'Supercomfort' cab mounted on rubber blocks, a hydraulically suspended seat and a three-speed heating and ventilation system.

119. The four-cylinder 70 hp Fiat 70-90 was one of eight new 90 series tractors launched in 1986.

120. The Fiat G 190 was made at Winnipeg on the same production line as the New Holland Ford 70 series.

badge on the bonnet when Fiat Geotech, the agricultural division of Fiat, merged with Ford New Holland in 1990. The new company, called NH Geotech, became New Holland Inc in 1993. The separate Ford and Fiat tractor ranges were retained for several years and the Fiatagri UK headquarters was still at Bury St Edmunds when the four-wheel drive Fiat F100 Winner range was introduced in 1991. Six-cylinder 98, 110, 118 and 130 hp Fiat Iveco 8000 series engines provided the power for the F100, F110, F120 and F130 Winner tractors with push-button controlled 540, 750 and 1,000 rpm power take-off shafts. Other features included electronic controls for front and rear hydraulic linkage and a thirty-two forward and sixteen reverse hi-lo gearbox. A sixteen-speed shuttle box, a sixty-four and thirty-two creep box or an economy twenty forward and sixteen reverse Eco Speed gearbox were optional extras. The F115 was added in 1993 and improved engines gave the Winner range increased torque back-up and reduced fuel consumption. Category II and III three-point linkage and a new air-sprung seat were added at the same time.

The two- and four-wheel drive Fiat 94 series, the 55-85 and 60-85 crawlers and the 62-86, 72-86 and 82-96

orchard tractors with 60, 70 and 80 hp engines appeared in 1993. The 65–85 hp Fiat 94 series designed with 'the profit-conscious farmer of the 1990s in mind' had three and four-cylinder Iveco engines, a synchronised reversing transmission system and a three-speed power take-off with ground speed. The Fiat 'Steering-o-Matic' full-drive system was a new feature on the 55 and 60 hp crawler tractors that were steered with a single joystick leaving the driver's other hand free to operate the hydraulic system and other controls.

New Holland was the dominant brand in the UK when the mechanically identical 170–240 hp blue Ford 70 series and terra cotta Fiat G series made on the same Winnipeg production line were launched in 1994. The Fiat G series was withdrawn from the UK market in the late 1990s.

Ford

The first Fordson Model F tractors were made at Dearborn in America in 1917 and at Cork in Ireland in 1919. The Model N replaced the F in 1929 and production was transferred to Dagenham in 1933.

121. *Three types of mudguard were used on the Fordson E27N during its six-year production run.*

Many thousands of Model N tractors had been built when the E27N Fordson Major superseded it in 1945. Henry Ford also built the Ford Ferguson 9N for Harry Ferguson in Detroit between 1939 and 1947 and the Ford 8N, which replaced the 9N, was made in America until 1952.

The Ford Ferguson 9N which cost $595 had an access panel in the bonnet similar to the later Ferguson FE35 for servicing the fuel tanks and radiator. Early 9Ns had a 10-gallon petrol tank and a 1-gallon reserve but later models were modified to run on paraffin. A shortage of materials as a result of America entering World War Two led to the introduction of the Ford Ferguson 2N in 1942. It was an economy version of the 9N on steel wheels with magneto ignition and a starting handle but when rubber became more plentiful the 2N was available on pneumatic tyres.

Henry Ford and Harry Ferguson parted company in 1946 and within a year the 9N had been replaced by the 35 hp cream and red Ford 8N with the Ferguson hydraulic system. The new tractor was similar to the 9N but with an extra forward gear and more than 500,000 had been made when it was discontinued in 1952. The 8N was the subject of a much publicised lawsuit, eventually won by Ferguson, concerning the use of his patented hydraulic system. Ford celebrated their 50th anniversary in 1953 with the launch of the 31 hp NAA Golden Jubilee tractor. Known as the Red Tiger it had a wheatsheaf emblem to commemorate the event.

The E27N Fordson Major (E for England, 27 hp, Model N) retained the Model N 27½ hp petrol and vaporising oil side-valve engines with cast-iron pistons, splash-feed lubrication, magneto ignition and starting handle. The three forward and one reverse gearbox was also used but the new tractor had a single-plate clutch. The crown wheel, differential and spur gear final drive transmission replaced the Model N's worm and wheel final drive. The E27N was available with a standard or optional high-speed transmission with top speeds of 4.17 and 7.48 mph respectively.

The Standard and Agricultural Rowcrop models of the dark blue and orange Fordson Major, introduced in 1945, had steel wheels, drum steering brakes, adjustable wheel track and a swinging drawbar. Electric starting, hydraulic lift and a side-mounted belt pulley were included in the list of optional extras – pneumatic tyres became available in 1946.

The Land Utility model of the Fordson Major on pneumatic tyres, but without steering brakes or an adjustable wheel track, cost £296 when it appeared in

1946. An industrial version of the Fordson Major and an optional factory-fitted 45 hp Perkins P6 diesel engine were added in 1948. County Commercial Cars and Roadless Traction Ltd used Fordson Major tractor skid units for their half-track and crawler conversions.

The Model N petrol/vaporising oil engine used for the Fordson Major was somewhat outdated in design and it was not long before Ford tractor dealers were busy with engine overhauls. The Fordson tractor engine exchange scheme was introduced in 1949 to help their dealers cope with the extra work and in the early days of the scheme the Ford Motor Co supplied a new block in exchange for the old one which was sent to Dagenham for reconditioning. The dealer overhauled the cylinder head, manifold, magneto and other engine parts. When the engine exchange scheme was established the worn cylinder block was exchanged for one that had been reconditioned at Dagenham. An advertisement for the tractor engine exchange plan for the Model N and E27N offered a replacement engine at the relatively low cost of £29 plus an extra charge to install the engine.

Farmers had the choice of a twin-lever Varley hydraulic lift system (page 139) with a separate control for external services or the single-lever Smith lift for the E27N. Both were self-contained bolt-on hydraulic

pump units driven by a shaft from the tractor gearbox. As they were unable to use the patented Ferguson design of hydraulic system with its built-in overload safety mechanism Ford offered an optional top link with an automatic overload release mechanism. A hydraulic damper in the top link was connected by a mechanical linkage to the clutch pedal and the increased forward thrust created in the top link when a mounted implement hit an obstruction actuated the linkage and disengaged the clutch. An adjustable relief valve in the hydraulic damper was used to vary the sensitivity of the release mechanism for different soil types and implements.

The Ford Motor Co bought a foundry and factory at Leamington Spa in 1945 to manufacture their own Ford Elite two- and three-furrow trailed ploughs, cultivators and ridgers. A mid-mounted toolbar made by the Martin Cultivator Co, Robot vegetable planters, Allman powder dusters, Compton loaders, Dening saw benches and Bamford mowers were some of the implements approved for use with the Fordson Major.

The New Fordson Major E1A, classed as a 40 hp tractor, was launched at the 1951 Royal Smithfield Show. This brand new tractor with an overhead-valve engine had a single-plate clutch, a six forward and two reverse gearbox with a top speed of 13 mph and spur gear final

122. The three-cylinder Fordson Dexta was launched in 1957.

123. The Fordson Super Major was the first 'Major' with position control, Qualitrol and flow control hydraulics.

drive. The same cylinder block was used for the petrol, vaporising oil and diesel engines with a pneumatic governor which were rated at 35, 34 and 37 hp belt hp respectively. Three-point linkage and power take-off were optional extras when the tractor was launched but they soon became standard equipment.

The New Fordson Major was used for various conversions including County full tracks, Roadless half tracks, the narrow KFD orchard tractor made for Kent Ford Dealers and a tricycle-wheeled Fordson Major made for the American market by Roadless Traction. There were two models of the KFD Fordson Major. The low-clearance KFD68 was 52 in high and 68 in wide but the KFD52 was only 52 in wide. A modified and more powerful Mk II Fordson Major with a 51.8 hp diesel or a 39½ hp petrol or 38½ hp vaporising oil engine was introduced at the 1956 Royal Smithfield Show. Other improvements included a dual clutch for live hydraulics and power take-off, power-assisted steering and a 'New-Comfort-Seat' with a seat cushion and backrest.

The Dexta, launched in 1957, had a three-cylinder 32 hp diesel engine with an in-line fuel injection pump and pneumatic governor. It was also the first Fordson tractor with a position and Qualitrol (draft control) hydraulic system. The standard specification included a six forward and two reverse gearbox, adjustable wheel track, category I three-point linkage, independent drum brakes, power take-off and a tip-back seat complete with a cushion. Optional extras included a dual clutch with live power take-off and hydraulics, an automatic pick-up hitch and power-adjusted rear wheels.

The gearbox and differential were strengthened on the 51.8 hp Fordson Power Major when it replaced the New Fordson Major in 1958. The hydraulic lift capacity was increased, rear wheel track adjustment was simplified and the new model had a hand throttle under the steering wheel. The vaporising oil-engined Power Major was discontinued in 1956 when most new tractors were sold with a diesel engine, however it was still possible to buy a Power Major with a petrol engine.

The Super Major replaced the Power Major in 1960. The engine was unchanged but the Super Major had disc brakes, a diff-lock and 'slimline' styling. The most significant improvement was the addition of the Qualitrol, position control and flow control hydraulic system. Qualitrol was the Fordson version of draft

124. A weather cab and Select-O-Speed transmission were optional for the Ford 5000.

control and flow control was used to limit oil flow to the hydraulic lift cylinder or external hydraulic ram. The linkage had quick-change category I and II balls and a category I top link end was included in the toolkit. The New Performance Super Major 53.7 hp diesel-engined tractor introduced in 1963 had grey wheels and mudguards.

A restyled 32 hp Dexta with an improved gearbox and hydraulic system appeared in 1960 when the Ford Motor Co introduced the petrol-engined Dexta for some overseas markets. The orange and blue colour scheme was still current when the 39½ hp Fordson Super Dexta was launched in 1961 but it was changed to a predominantly blue livery with white wheels and mudguards in 1962. The new colour scheme was also used for the Super Major and Super Dexta when engine power was increased to 53.7 and 44½ hp respectively but the Dexta remained unchanged at 32 hp.

Tractor production at Dagenham came to an end in 1964 when Ford introduced the new 6X or 1000 series tractors made for worldwide markets at Antwerp, Highland Park in America and a new factory at Basildon in Essex. The Dexta 2000, Super Dexta 3000

and Major 4000 with three-cylinder direct injection Ford engines rated at 37, 46 and 55 hp and the four-cylinder 65 hp Super Major 5000 cost £680, £765, £885 and £965 respectively. The 2000 was a basic tractor with a single-plate clutch, six forward and two reverse gears, drum brakes and category I hydraulic linkage. Independent power take-off, eight forward and two reverse gears, double-acting top link draft control hydraulics and diff-lock were standard on the larger models. Drum brakes were retained for the 3000 but the 4000 and 5000 had wet disc brakes. There was an optional weather cab and the Select-O-Speed transmission with four creeper, four field and two road speeds was available for the 3000, 4000 and 5000. A single lever was used to select the required gear in the ten forward and two reverse speeds in the clutchless change-on-the-move Select-O-Speed gearbox. Epicyclic gear units and band brakes engaged the selected gear and an inching pedal provided precise control when manoeuvring in confined spaces or hitching implements. Some drivers called it the 'jerk-o-matic' gearbox as there could be a pronounced jump in forward speed when changing up to a higher group of gears or vice versa.

The Dexta and Major names were dropped from the 1000 series tractors in 1966 when the Ford Force 6Y Series appeared in 1968 and engine power for the 2000, 3000, 4000 and 5000 was increased to 39, 47, 62 and 75 hp respectively. A Ford Force 5000 skid unit was used for the Italian-built Mailam 5001 crawler (page 85) and the Doe 5100 made in 1971 with a six-cylinder Ford industrial engine.

The 94 hp Ford 7000, introduced in 1971, followed the general design of the 5000 with eight forward and two reverse gears, independent power take-off and power-assisted steering. The 7000 was the first Ford tractor with a factory-fitted turbocharger.

The old Ford Force 6Y cab was used for the first production models of the new 600 series launched late in 1975. Within a few months a new luxury two-door quiet cab advertised 'as spacious, quiet and comfortable as a car' was standard on the seven-model 600 series ranging from the 37 hp 2600 to the 97 hp 7600. Ford changed from a dynamo to an alternator charging system to provide the extra power required for the improved lighting and in-cab controls. The 5600, 6600 and 7600 were also the first Ford tractors with the option of Load Monitor hydraulics. Load Monitor was an alternative form of draft control which measured variations in torque in the drive shaft between the gearbox and the crown wheel. The system transmitted signals to the hydraulic control valve which raised or lowered the lift arms according to need.

The six-cylinder 126 hp Ford 8600 and 148 hp 9600 originally launched in the UK and America in 1972 completed the Ford 600 series. Both tractors had a sixteen forward and four reverse Dual Power transmission, hydrostatic steering, hydraulic disc brakes, lower link sensing hydraulics and a 540/1000 rpm power take-off.

Four 700 series models launched in 1976 with new quiet cabs, restyled exhaust pipes and air cleaner intakes gave Ford a range of eleven two-wheel drive 600 and 700 series tractors in the 37–153 hp bracket. The flat floor cab on the 6700, 7700, 8700 and 9700 was a sign of things to come with the pendant pedals and driving controls grouped on the right-hand side of the seat to provide an uncluttered floor. The 6600 and 7600 remained in production but the 8600 and 9600 were replaced by the 128 and 153 hp two- and four-wheel 8700 and 9700 with six-cylinder engines.

The Ford FW30 265 hp articulated tractor built by Steiger in America made its UK debut at the 1978 Power in Action demonstration held in Suffolk. The new FW (Four Wheel drive) tractor had a Cummins

125. Ford FW tractors were made by Steiger in America.

126. The four-wheel drive Ford TW 15 was launched in 1983.

V8 engine, a twenty forward and four reverse constant mesh gearbox and category III hydraulic linkage. The FW60 with a 335 hp Cummins engine and a twenty forward and four reverse constant mesh gearbox and category III hydraulic linkage was launched at the 1980 Royal Smithfield Show. The FW60 was also sold in the UK but the 210 hp FW20 and the 295 hp FW40 were not. A modified FW60 with a 325 hp Cummins engine appeared in 1984 and an optional ten-speed automatic gearbox for the FW60 was added in 1985. Production of Ford FW tractors ceased when Case IH acquired Steiger in 1987.

The Ford TW series was introduced to European farmers at the 1979 SIMA Show in Paris. The Ford TW range was made in two- and four-wheel drive format but most of the TW10, TW20 and TW30 tractors sold in Britain were four-wheel drive models. The 128, 153 and 188 hp tractors had a sixteen forward and four reverse Dual Power transmission, lower link sensing hydraulics and self-adjusting brakes.

To compete with a growing range of mainly Japanese-built compact tractors on the UK market Ford introduced the twin-cylinder diesel-engined Ford 1000 made by Shibaura in Japan in 1973. The Ford 1000 had a three-range gearbox with three forward gears and one reverse gear in each range, 23 hp at the power take-off and adjustable wheel track. The Ford

1100, 1200 and 1900 compact models, also made by Shibaura, were added in 1980. The two-wheel drive 13 hp 1100 and the two- and four-wheel drive 16 hp 1200 had ten forward and two reverse gears. The two- and four-wheel drive 28 hp Ford 1900 had a twelve forward and four reverse gearbox.

Ford made their 5,000,000th tractor in 1981 and introduced the first of eleven new Ford 10 series tractors in the same year. Based on the earlier 600 and 700 series the restyled three- and four-cylinder 10 series tractors ranged from the 44 hp 2610 to the 98 hp 7610 and 7710. The four-wheel drive six-cylinder 116 hp 8210 launched in 1982 superseded the earlier two- and four-wheel drive 115 hp 8100 and four-wheel drive 8200. The more significant changes on the 10 series included an eight forward and four reverse change-on-the-move Synchroshift gearbox with optional Dual Power, an improved hydraulic system and an average 10 per cent power increase across the range compared with the previous 600 and 700 series tractors. Four-wheel drive with a central drive shaft and an automatic diff-lock with a switch-operated on-the-move drive engagement was a factory-fitted option for the 50 hp 4110 upwards. The choice of a low-profile (LP), all-purpose (AP) or de luxe quiet cab was one of the options for the Ford 10 series tractors.

More new models were launched in 1983 when the TW15, 25 and 35 with 132, 154 and 186 hp six-cylinder engines superseded the TW10, TW20 and TW30. The two smaller TW tractors had a 540/1,000 rpm power take-off and self-adjusting disc brakes. A 1,000 rpm power take-off and multiple disc brakes were standard on the TW35. At the opposite end of the power scale the 16 hp Ford 1210 and 32 hp 1910 compact models replaced the 1100, 1200 and 1900 the same year. An optional hydrostatic transmission for the 1210 and the new two and four-wheel drive 26 hp 1710 were added in 1984.

The Ford Force II tractors with Super Q luxury cabs on models from the 6610 to the TW35 appeared in 1985. A new-style low roofline cab and forward-facing exhaust were the more obvious changes but there was also a more efficient alternator and telescopic stabilisers replaced the earlier chain and turnbuckle design. Hydrostatic steering was standard on the Force II 5610, 6610 and 7610.

127. The 116 hp Ford 8210 replaced the 8200 in 1982.

Ford Tractor Operations bought the farm machinery division of New Holland from the Sperry Corporation in 1986 for a sum in excess of $250m and the new company became Ford New Holland. The Basildon factory celebrated twenty-five years of tractor production in 1989 when a batch of 7810 models had a silver colour scheme. The jubilee was also marked with a special offer of two tractors for the price of one – but the second was a small pewter replica of the real thing!

The Ford 20 series compact tractors replaced the 10 series when they were launched at the 1988 Royal Smithfield Show. Built by Shibaura in Japan the 20 series consisted of the 1120, 1220, 1520 and 1720 with 15–28 hp three-cylinder engines together with the four-cylinder 1920 and 2120 rated at 33 and 41 hp. The 1120 and 1520 were discontinued in 1993, the 1220 and 1720 were dropped in 1996 but the Ford 1920 and 2120 remained in production until 1998.

The earlier 132, 154 and 186 hp TW series engines were used for the 8630, 8730 and 8830 Ford 30 series

launched in 1989. They had an eighteen forward and nine reverse Funk Powershift transmission and improved hydrostatic steering. The Powershift gearbox, with single lever clutchless gear changing and shuttle reverse, was monitored by a microprocessor designed to prevent engine or transmission overload due to incorrect gear selection. The smaller two- and four-wheel drive 30 series were multi-purpose tractors suitable for arable and stock farms. Options included two designs of cab, four types of transmission and a standard or economy 540 rpm power take-off at 1,750 engine rpm or the more economical engine speed of 1,450 rpm. A buzzer in the cab warned the driver when the power take-off exceeded 630 rpm in the economy setting. The Ford Performance Monitor with a radar unit measuring ground speed was standard on the 8630, 8730 and 8830 tractors. The 30 series was extended later in the year when the two- and four-wheel drive 51 hp Ford 3930 and the 61 hp 4630 replaced the 4110 and 4610. Both models had an eight-speed Synchro Shuttle gearbox or optional sixteen

128. The Ford New Holland 7840 was launched in 1991.

forward and eight reverse Dual Power transmission and a flat floor cab. The 70 hp 5030 was added to the range in 1992.

Ford New Holland merged with Fiat Geotech in 1991 with the New Holland Geotech headquarters at Brentford in Middlesex and it was announced that both Ford and Fiat tractor ranges would be made at Basildon. This proved to be true when the blue Ford 40 Series SL and SLE tractors and the Fiat Winner range, launched at the 1991 Royal Smithfield Show. The Ford 40 series was advertised as the first completely new tractors made at Basildon for twenty-seven years. There were to be six two- and four-wheel drive Ford 40 series tractors with 75–120 hp PowerStar engines, new transmissions, hydrostatic steering and SuperLux cabs. The 75–100 hp SL models had a twelve forward and reverse SynchroShift gearbox with reverse shuttle. The SLE version had the latest Electro-Shift transmission with sixteen speeds in forward and reverse. The higher SLE specification also included the Ford 'ElectroLink' closed centre lower-link sensing hydraulics and an Electro-Command electronic instrument panel in the flat floor cab with a passenger seat. Front power take-off and three-point linkage were optional extras.

The company name was changed to New Holland in 1993 when the turbocharged 125 hp Ford 8340 was added to the 40 Series. A twenty-four-speed Dual Power transmission with an 18 per cent speed reduction in

each gear was added to the list of optional extras for SL models. The 70 series tractors with either Ford New Holland blue or Fiat terra cotta colour schemes were launched in 1994. The Ford New Holland 8670, 8770, 8870 and 8970 with maximum engine outputs of 170–240 hp were, apart from their colour, identical to the Fiat G170, G190, G210 and G240. Similarly, the blue New Holland Ford 60 series, introduced in 1996, was identical to the Fiat M series. The four 60 series models with six-cylinder engines comprised of the naturally aspirated 100 hp 8160/M100 and 115hp 8260/M115 and the turbocharged 135 hp 8360/M135 and 160 hp 8560/M160. Various options included Shuttle Command, Dual Command or Range Command transmissions and mechanical or electronic hydraulic controls.

The four-cylinder TS90, TS100 and TS110 introduced in 1997 and the six-cylinder TS115 added in 1998 were the last tractors with New Holland Ford decals on the bonnet. The blue Ford oval badge and the Fiatagri name finally disappeared from the tractor scene in favour of the New Holland logo in November 1995.

Fowler

John Fowler & Co at Leeds, a leading steam traction engine manufacturer in the mid-1800s, were involved in the production of the Wyles Motor plough in 1912 and introduced the Fowler Gyrotiller with a 225 hp

ESSENTIAL READING FOR STEAM ENTHUSIASTS!

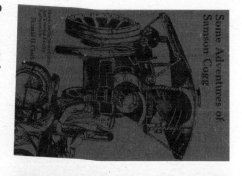

Some Adventures of
Samuel Cogg

We have some copies of

Memories of Steam
Rallying

Paddle Steamers at
War 1914-18

Traction Engines

MASHAM CELEBRATES 50 YEARS IN STYLE!

John Maund reports from this year's special commemorative rally at Masham in Yorkshire.

B ack in September 1965, a small steam rally was held in the North Yorkshire town of Masham. The individuals who organised and held that event did so to try and raise money to save the local attractive town hall from ruin, but the event itself suffered badly from the weather and very little money was raised. However, that was only the first of what was to become a very well

129. The 40 hp Fowler VFA crawler was based on the Field Marshall.

Ricardo petrol engine in 1927. Later Gyrotillers had a 150 or 170 hp MAN diesel engine. Production continued until 1935 when the 170 hp model cost £6,000.

Fowlers turned their attention to building crawler tractors in the early 1930s when they introduced the 25 drawbar hp 25, the 30 hp 3/30, and 40 hp 4/40. The 25 had a three-cylinder Fowler-Cooper engine, a single-plate clutch and the 30, 40 and 80 hp crawlers had Fowler-Sanders diesel engines. Optional equipment included a belt pulley, power take-off and a Gyrotiller attachment.

The Ministry of Supply took control of John Fowler & Co from 1941 when battle tanks were made at the Leeds factory. Rotary Hoes Ltd, with plans to build a range of FD crawler tractors, acquired the company in 1945 and the FD2 with a drive shaft for a Rotavator appeared in 1946. However, within a year Rotary Hoes sold the business to the TW Ward Group and production of the 24 hp Fowler FD2, along with the 35 hp FD3 and 54 hp FD4 crawler tractors continued at Gainsborough. The FD3 had a three-cylinder Fowler B series engine, a hand-lever operated clutch, six forward and two reverse gears, a top speed of 4½ mph and clutch and brake steering.

The 40 hp Fowler Mark VF crawler, based on the single-cylinder Field Marshall, replaced the FD3 in 1948. The Mark VF, which cost £1,060 ex works, had

the same six forward and two reverse speed gearbox, controlled differential steering and a side-mounted belt pulley. Optional equipment included a power take-off on the left side of the crawler, a central power take-off chain driven from the side power shaft, electric lighting and street plates for taking the crawler on the public highway. The improved Fowler Mark VFA, based on the Series IIIA Field Marshall, replaced the Mark VF in 1952 and like the Series IIIA wheeled tractor most VFA crawlers had orange paintwork.

The Marshall Organisation made Challenger crawlers in the 1950s. The Challenger Mk II, the first of the four-model range, appeared in 1950. This was followed in December 1951 by the smaller Challenger 1, which cost £2,300. The Challenger 3 and 4 completed the mid-1950s range of Fowler Crawlers. The Challenger 1 had an unusual 50 hp twin-cylinder two-stroke Marshall ED5 diesel engine with five inlet and three exhaust ports for each cylinder. The pressure lubricated 5¼ in bore and 6 in stroke engine had a chain-driven blower and a centrifugal pump circulated the water in the cooling system. The diesel engine was started with a twin-cylinder water-cooled Coventry Victor donkey engine with a hand-lever operated clutch to engage drive to the diesel engine. The donkey petrol engine, started with a rope and later by a 24-volt electric starter, was left to run until the cooling system water, which also circulated through the ED5 diesel engine, had warmed it sufficiently to make it easier to start. There were four optional power take-off shafts for the Challenger I. They included the rear power take-off, a power shaft on both sides of the engine with one of them chain driven from the fan pulley and the front power take-off shaft. An extension to the crankshaft was used to drive the hydraulic pump.

The 80 bhp Challenger Mk II had an FD8 two-stroke diesel engine with 24-volt electric starting and six forward speeds with the first four instantly reversible with a separate forward/reverse lever. The Challenger III, later designated the Challenger 3, with the choice of a 105 bhp Meadows or a Leyland engine had a single-plate clutch, six forward and four reverse gears and clutch and brake steering. The 150 bhp Challenger 4 with a six-cylinder Meadows diesel under the bonnet also had six forward gears and four in reverse, an independent rear 1,000 rpm power take-off shaft and a front power take-off which ran continuously at engine speed.

The Challenger Mk II with a Leyland engine became the Challenger 22 in 1958 when it was restyled to match the wheeled Marshall MP6 and Track Marshall crawlers. Other changes included a 100 rpm increase in engine speed, the fuel tank was relocated behind the seat and the gear ratios were changed to give a 20 per cent lower speed in each gear.

The Thomas Ward Group bought Marshall Sons & Co at Gainsborough in 1935 and John Fowler & Co at Leeds joined the group in 1946. The Fowler factory at Leeds was closed in 1974 but tractor production continued at Gainsborough. British Leyland bought the company in 1975.

Fraser

Fraser Tractors of Acton in London made a brief appearance on the agricultural scene in the early 1950s with the 45 or 60 hp Fraser crawler. Sales

130. The Fowler Challenger I had a twin-cylinder two-stroke Marshall diesel engine.

literature at the time explained that the tractor could have a petrol or vaporising oil engine of a well-known make, or a 45 hp Petter or 65 hp Russell Newbery diesel engine to suit customer requirements. The engine was started with either a 24-volt electric starter motor or a wind-up inertia starter. The Fraser had an automatic centrifugal clutch, four forward gears and one reverse, hand-lever controlled multi-disc steering clutches and cast tracks. A power take-off and a belt pulley were optional extras. An advertisement explained that the extreme simplicity of its controls was a unique feature of the Fraser crawler as there were no foot pedals and by simply opening up the throttle the centrifugal clutch engaged the drive to the tracks. Returning the engine to tick-over speed automatically disengaged the clutch. It was also pointed out that the automatic clutch enabled the driver to move the tractor in either direction while standing on the ground to hitch an implement. This would undoubtedly be a modern farm safety inspector's nightmare!

The Fraser cost £1,375 when it was introduced in May 1950 but the power take-off, belt pulley and winch were optional extras. Potential purchasers were advised that ample seating space was provided for the driver and if necessary there was room for two

131. The Fraser crawler was exhibited at the 1950 Royal Smithfield Show.

people to sit comfortably on the tractor. Fuel and maintenance costs were said to be very low and the tank held enough fuel for one-and-a-half-day's work.

Garner

Motor dealer Henry Garner at Birmingham, who imported tractors from America during World War One, entered a Garner tractor with a 30 hp paraffin engine at the 1919 Lincoln tractor trials. Compared with the Fordson N the tractor was expensive and so Garner turned his attention to manufacturing lorries. A subsidiary company, Garner Mobile Equipment Ltd of London, was established in 1947 to manufacture Garner pedestrian-controlled tractors. The four-wheel Garner light tractor, developed from the earlier two-wheel model, appeared in 1949. Garner was one of

132. Garner light tractors had a top speed of 10 mph.

several companies making small four-wheeled tractors in the late 1940s for smallholders and market gardeners. The many who used a horse or a two-wheeled garden tractor appreciated the advantage of riding on a tractor which, unlike a horse, did not have to be fed and watered every day of the year.

The four-wheel Garner light tractor had the same 6 hp JAP Model 5 air-cooled petrol engine and centrifugal clutch as the two-wheel model and the specification included a three forward and one reverse gearbox, differential, steering brakes and a roller chain drive to the rear wheels. The driving seat was positioned in front of the rear-mounted engine and transmission and the driver had to reach behind his back to change gear. The tractor cost £197.15s on steel wheels, pneumatic tyres adding £5 to the price. Other extras included a belt pulley, power take-off, a plough and various toolbar attachments. The Garner had a drawbar for trailed implements and a hand lever was used to raise and lower the mid- and rear-mounted toolbars and the single-furrow plough.

Sales literature advised potential buyers that the 'Garner is the leader of light tractors which really can plough'. However, as it was rather short of power when the going was tough a slightly wider and longer Garner with a 6 or 7 hp JAP Model 6 air-cooled petrol or vaporising oil engine appeared in 1950. The new model, which could be used with a mid-mounted reversible plough, cost £269.15s. A haulage model of the Garner was exhibited at the 1953 Royal Smithfield Show and some, with either a 7 or 10 hp JAP 55 twin petrol engine and a top speed of 8 mph, were used by the London Docks and Inland Waterways Executive. The Garner light tractor was too expensive when compared with a Ferguson TE20 and the last ones were made in 1955.

133. Introduced in 1948 the Gunsmith light tractor could plough two acres or more in

Gunsmith

Farm Facilities Ltd at Maidenhead made the three-wheel Mk I and Mk II Gunsmith light tractors between 1948 and 1952. Designed by Frederick Gunn and Harold Smith, the Gunsmith had the same 6 hp air-cooled JAP engine as the pedestrian-controlled BMB Plow-Mate garden tractor made by Brockhouse Engineering. A few Gunsmith tractors had an alternative Briggs & Stratton model ZZ engine. The BMB Plow-Mate's two forward and one reverse gearbox and rear axle were also used for the Gunsmith. The gearbox had top speeds of 2 and 3 mph, which by interchanging the engine and gearbox shaft pulleys, were increased to 3 and 5 mph. A tiller handle was used to steer the single front wheel and a foot pedal controlled an over-centre tensioner pulley which served as a clutch to engage the flat belt drive from the engine to the gearbox. A sales leaflet explained that by 'fitting a new belt the clutch could be relined in a matter of seconds with no mechanical knowledge required'. Implements could be towed from the drawbar or attached to mid- and rear-mounted tool frames raised and lowered with a hand lever.

The Mk II Gunsmith, which cost £197.10s when it made its debut at the 1950 Royal Smithfield Show, had solid rear wheel centres instead of the spoked wheels of the earlier model. A steering wheel and steering box replaced the tiller handle and an improved transmission with a system of multiple vee-belt pulleys on the engine and gearbox provided four forward and two reverse gears with a top speed of 8 mph. FarFac accessories, also made by Farm Facilities for the

Gunsmith, included a disc harrow, cutter-bar mower and a hammer mill.

Sales literature claimed that the Mk II Gunsmith would 'work as well as any big tractor, often more precisely and always at less cost. Furthermore youngsters will find it so easy to operate they will delight in doing men's work.' It was also suggested that, 'After the Gunsmith has been tried out and you have seen how much work can be done with this amazing little tractor you won't rest until you own one.'

Many of the 300 Gunsmith light tractors made during its six-year production run were exported to Australia, Canada and South America. One owner, writing a testimonial letter in praise of the Gunsmith, described it as 'a wonderful machine that would do five hours' work in one hour while he sat in an armchair' – his description of the Gunsmith's seat. Garden Machinery Ltd made a four-wheel version of the Gunsmith when they had purchased the remaining stock of Gunsmith parts in 1952.

Chapter 4

Hanomag – John Deere

Hanomag

Hanomag, or Hannoveresche Maschineblau AG, were making steam engines in the 1840s, six-furrow motor ploughs in 1912 and eight-furrow 80 hp motor ploughs by 1919. The first 25 hp Hanomag Z25 crawler was made in 1925 and this was followed by the Z50 crawler with an 8-litre paraffin engine in the late 1920s. The 40 hp D52 with a four-cylinder four-stroke engine, introduced in 1930, was the first wheeled Hanomag diesel-engined tractor. The lightweight wheeled RL25 with the choice of a three- or four-speed gearbox appeared in 1937 and

HANOMAG HANNOVER Germany

K90, K55

R55, R45

R35, R27

R24

R12

HANOMAG's most modern machinery aids farming all over the world

The HANOMAG production programme includes wheeled tractors ranging from 12 to 55 HP, and tracklayers of 55 and 90 HP. Perfectly matched mounted and trailed implements are available for all HANOMAG tractor types.

Please inform yourself on HANOMAG tractors, without obligation, by contacting your nearest HANOMAG dealer, or writing direct to Department "VW", HANOMAG Works, Hannover/W. Germany. We will be pleased to advise you on any specific problem.

134. The 1956 Hanomag tractor range.

an improved R25, also with the option of a three or a four forward speed gearbox and three-point hydraulic linkage, replaced the earlier model in 1949.

The Hanomag K50 crawler made between 1933 and 1941 was relaunched in 1948 as the KV50 with a 50 hp four-cylinder Hanomag diesel engine, a three forward and one reverse gearbox, controlled differential steering with a steering wheel and the choice of standard and narrow track centres. The K55 with a four-cylinder engine, three forward gears and one reverse and controlled differential steering replaced the KV50 in 1951. HL Arnes & Co, the Hanomag concessionaires for the UK, introduced the red-painted Hanomag R22, R25 and R45 wheeled tractors to British farmers at the 1950 Royal Smithfield Show.

The 1952 Hanomag price list included 16–55 hp wheeled tractors and 55–90 hp crawlers with four-stroke diesel engines. The wheeled models included the three-cylinder 22 hp R22 with the option of five or eight forward gears and the four-cylinder 45 hp Hanomag R45. The single-cylinder 12 hp two-stroke R12 with a dual seat was introduced in 1953 and an improved model with a rounded bonnet and single driving seat appeared in 1954. Other mid-1950s Hanomag wheeled tractors included the R16 with a twin-cylinder four-stroke engine, the 24 hp twin-cylinder R24 and the 35 hp four-cylinder R35, both with two-stroke engines.

Hanomag tractors were a rare sight on British farms in the early 1950s but a few, including the K90 crawler launched late in 1952, were sold in the UK. The K90 had a five forward and four reverse gearbox, hand lever-controlled clutch and brake steering and the model number indicated that it had a six-cylinder Hanomag 90 hp diesel engine. The K90E and K90L were variants for use with a

135. Sales literature explained that the Hanomag Perfekt 300 had an 'ideal upholstered parallelogram seat with hydraulic suspension adjustable to the height and weight of the driver'.

bulldozer blade and front loader. The smaller Hanomag K60 with a 60 hp twin-cylinder blower-assisted two-stroke diesel engine and a six forward and three reverse transmission was added in 1956. Hanomag also introduced the R18 wheeled tractor for the German market and a tool carrier with an 18 hp twin-cylinder two-stroke diesel engine in the same year.

The K65, which superseded the K60 in 1958, was made for two years. Hanomag Crawler Tractors Ltd exhibited the K320 and the K90 at the 1960 Royal Smithfield Show. The K320, with the same engine as the earlier K65, had a 24-volt electrical system and a six forward and three reverse gearbox. Early 1960s Hanomag wheeled tractors with four-stroke diesel engines included the 25 hp Perfekt 300, 32 hp Perfekt 400, and the 38 hp Granit 500. The 50 hp Brilliant 600 and 60 hp R460 completed the range. The Granit 500 had a direct injection diesel engine with the choice of a nine or twelve forward speed transmission and optional front-wheel drive. The six-cylinder Robust 900 with twelve forward and three reverse gears was added in 1967. Introduced in 1970 the 75 hp Hanomag Brilliant 701 with the same twelve-speed gearbox as the Robust 900 was one of the last tractors built at Hanover.

Due to disappointing sales figures Hanomag tractor production came to an end in 1971.

Howard

Australian engineer AC Howard introduced the 22 hp Howard DH22 tractor for use with his rotary cultivator in 1928. The tractor was successful in Australia and was made there for nearly twenty years but little interest was shown when it was exhibited at the Royal Agricultural Show of England in the early 1930s.

AC Howard, who formed Rotary Hoes Ltd at East Horndon in Essex in 1938, bought John Fowler & Co at Leeds in 1945. Rotary Hoes used Fowler's expertise in crawler tractors to design the FD2 crawler tracklayer with a drive shaft for the Howard Rotavator which was launched in 1946. Within a year or so Rotary Hoes made a handsome profit when they sold John Fowler & Co to the TW Ward Group at Gainsborough in Lincolnshire.

Marshalls made the FD2 crawler at Gainsborough for the next year or so and meanwhile Rotary Hoes used the basic design of the FD2 for the new Howard Platypus 28 crawler tractor. About twenty Platypus 28 tractors with a 28 hp Standard petrol engine, six forward and two reverse gears, controlled differential steering, three

136. Introduced in 1952 the 35 hp Howard Platypus 30 had a Perkins P4 (TA) diesel engine.

power take-off shafts and hydraulic linkage were made by the Platypus Tractor Co, a subsidiary of Rotary Hoes at East Horndon. Platypus production was transferred to a factory at Basildon in 1952 and a diesel-engined version of the Platypus 28 appeared in the same year.

The Platypus 30 crawler was introduced in 1952 and many of the 400 or so tractors made during its five-year production run were exported, mainly to sugar cane growing areas. The Platypus 30 had a 35 hp Perkins P4 (TA) indirect injection engine although a few were sold with a 28 hp four-cylinder petrol engine. The Platypus 30 had six forward and two reverse gears and controlled differential steering and was advertised as a tractor which could 'pull a four-furrow plough 9 in deep in third gear'. It had a 540 rpm rear power take-off and a two-speed power shaft on each side of the tractor. One of the side shafts was used with an optional belt pulley and the other could be used to drive a Howard Rotavator.

The 1955 narrow-tracked Platypus PD2 for rowcrop work was similar to the earlier Platypus 30. The PD2 tractor was also used with an angle dozer blade and with

Howard Bulloaders, Bogmasters and Bogwagons. The Howard Bulloader was an overloader shovel with a bucket which picked up spoil in front of the tractor and a pair of double acting rams lifted the bucket over the top of the tractor and discharged the spoil on to a heap or into a trailer behind the machine. Sales literature explained that the Bulloader, with a canopy to protect the driver from falling debris, only took thirty seconds to complete one cycle of digging, collecting and discharging a loaded bucket. The PD2 Bogmaster with 36 in wide tracks was made for the Irish peat bogs and a few PD2 Bogwagons with a tipping load platform were sold in Japan.

The Platypus 50 launched in 1954 and the PD40 added in 1955 had 51 hp Perkins engines. The 50 had a two-range six forward and four reverse gearbox and the PD40 specification included a six forward and two reverse gearbox, a two-speed power take-off and the option of 16 or 24 in wide track plates. A couple of Platypus tractors were made with Leyland diesel engines and another had a 120 hp engine and a thirty-four speed gearbox. The Platypus proved unprofitable and the last one was sold in 1957. Trailed Rotavator production continued at the Basildon factory for a couple of years and was then transferred to East Horndon.

Hürlimann

Hans Hürlimann made an 8 hp tractor with a single-cylinder engine at Wil near Zurich in 1929 followed by a 10 hp model in 1930. Both tractors had a petrol engine and a three forward and one reverse speed gearbox. The Swiss engineer, who was a pioneer of the direct diesel engine, introduced the much larger four-cylinder 45 hp Hürlimann 4DB85 with this type of diesel engine and an optional mid-mounted cutter bar in 1939. The Hürlimann D series with 45–85 hp engines was introduced in 1946. The 45 hp D100 with a diesel engine had hand and foot throttles, five forward gears and one reverse, a two-speed belt pulley, power take-off and a diff-lock. The orange-painted 28–40 hp H10, H12, H17 and H19 made between 1951 and 1955 were the last petrol-engined Hürlimann tractors but small-scale production of diesel-engined models continued at Wil for the next twenty years. The D150 and D200 launched in 1969 had a twelve forward and six reverse synchromesh gearbox and an independent power take-off. The 155 hp D155 added in 1971 was the first turbocharged Hürlimann tractor and an optional hydrostatic transmission was available from 1974. Hürlimann tractors were hand built to a high standard but strong competition for sales in the early 1970s caused the company serious financial difficulties. The situation improved in 1975 when Hürlimann became the Lamborghini tractor distributor in Switzerland. Early discussions with SAME came to nothing but the Italian tractor manufacturer, which already owned Lamborghini, bought Hürlimann Traktoren AG in 1975 and the group became SAME-Lamborghini-Hürlimann in 1979. A new range of Hürlimann tractors, including the H480,

H490, H510, H6130 and H6160 with a new light green colour scheme, appeared in 1979 with the H360 and H470 being added later in the year. The turbocharged Hürlimann engines, made in Switzerland for the H480/490 and the H6130/6160, were also used for some SAME and Lamborghini tractors built in Italy. At least twenty Hürlimann models in the 47–165 hp bracket were current in the mid-1980s. Most had Swiss-built power units but the Hürlimann H345, H355 and H355F introduced in 1982 had three-cylinder SAME engines.

WR Bridgeman & Son at Newbury imported a limited range of Hürlimann tractors in 1980 and the Berkshire company introduced the two-and four-wheel drive 110 hp H5110 with a twelve forward and three reverse synchromesh gearbox and optional twenty-four speed creep box at the 1980 Royal Show. An advertisement for the five-cylinder Hürlimann H5110 with lower link hydraulic sensing explained that the tractor had 'the power of six cylinders with the economy of four' and that the tractor's 'five cylinders run as smoothly as six and as cheaply as four'. Other Hürlimann tractors sold

137. The light green colour scheme of the 145 hp Hürlimann H-6130 replaced the earlier Hürlimann red paintwork in 1979.

by WR Bridgeman & Son at the time included the two- and four-wheel drive H480 and H490 with 82 and 90 hp Hürlimann engines and more powerful four-wheel drive models including the 130 hp H6130, 160 hp H6160 and the 200 hp H6200. Launched at the 1980 Royal Smithfield Show the six-cylinder H6200 had an electrically controlled fluid clutch that allowed all twelve forward speeds either to be decreased by 20 per cent or increased by 15 per cent.

Within a couple of years WR Bridgeman & Sons were importing two- and four-wheel drive Hürlimann tractors in the 65–162 hp

138. The Hürlimann XT-908 had a 100 hp six-cylinder engine with electronic injection control.

bracket. The 70 hp H470 was typical of the range. It had a direct injection engine with four individual water-cooled cylinders, a dual clutch and a twelve-speed synchromesh gearbox with three reverse gears. The standard specification was impressive with a dual-speed independent power take-off, hydrostatic steering, radial tyres and an air-conditioned safety cab with tinted glass and two-speed windscreen wipers. SAME (UK) at Thirsk in Yorkshire took control of Hürlimann tractor distribution in 1983 when the H480, H490 and H5110 rated at 82, 95 and 115 hp were advertised in the farming press. Hürlimann tractors were given a makeover in 1985 and the price list for 1986 included six restyled models from the 62 hp H360 to the turbocharged H-6170T with a 165 hp six-cylinder engine and a twenty-four speed gearbox. The tractors were relatively expensive and although sales were limited they remained on the British market until well into the mid-1980s.

After a gap of nearly ten years the light green tractors reappeared at the 1995 Royal Show when Motokov UK Ltd, who imported Zetor tractors, were appointed Hürlimann distributors. The Swiss name was the main link with the past as far as the Italian-built 85–190 hp four-wheel drive Hürlimann tractors were concerned.

They shared many components with the SAME and Lamborghini marques but with a higher specification which included a sophisticated power-shift transmission.

The Hürlimann XT, Elite and Master ranges were advertised in 1997 as 'quality tractors with a long Swiss pedigree made in Italy'. The four 85–105 hp XT models had a sixty-speed forward and reverse transmission while the 115 and 135 hp Elite XB and the Master range with 165 and 190 hp engines had a fifty-four speed forward and reverse electro-hydraulic transmission. An air-conditioned cab, electronic load-sensing hydraulics, four-wheel hydraulic brakes and a trailer braking system were standard equipment. Optional extras included a computerised performance monitor with a radar speed sensor and a remote control rear-view telecamera and screen. Motokov UK relinquished their interest in Hürlimann tractors in 1998.

Hydraulic Systems

Most farms had at least one tractor with hydraulic three-point linkage in the late 1940s. Some had a built-in hydraulic system while others had a bolt-on hydraulic unit with the pump driven from the tractor gearbox. Examples of bolt-on hydraulic units included the Smith lift and the Varley lift for the E27N Fordson Major and the International Harvester Lift-All system for Farmall tractors.

The Ferguson draft control and weight transfer system was so well protected by patents that other manufacturers were forced to use a basic lift and drop hydraulic system which required wheels to control the working height or depth of an implement. This prompted engineers to design other ways of controlling implement depth and transfer some of its weight on to the tractor. The Bulwark Cantilever traction attachment made by Salopian Engineers in 1954 was claimed to reduce wheelslip and to save fuel, time and labour by transferring implement weight onto the tractor. An advertisement of the time explained that the Bulwark Cantilever, designed for the Nuffield Universal, 'only costs £30, can be fitted in thirty minutes and will allow the driver to keep on ploughing, even in the wettest weather'.

An adjustable metal block on the rear axle housing supported the lower left-hand lift arm and with the depth wheel fully raised some of the implement's weight was carried on the tractor. A small hydraulic ram connected to the external services system on the tractor was used to adjust the height of the metal block and vary the working depth of the implement.

139. The Varley bolt-on hydraulic lift pump for the E27N Fordson Major was shaft driven from the tractor gearbox.

TCU (Traction Control Unit), also introduced in 1954, was the David Brown solution to wheelslip. The Meltham publicity department called wheelslip the 'invisible enemy' which they likened to 'the extra time it takes a fly to crawl up a wet window compared with the time spent crawling up a dry one'. They explained that the fly's problem was 'much the same as a tractor working in a wet field' but unlike the poor fly, David Brown owners would be able to overcome wheelslip with TCU. A depth wheel was still required for the implement when TCU was used to transfer some of its

140. The tractor hydraulic system was used to supply oil to the three-point linkage rams on the Doe tool carrier.

weight on to the tractor to overcome wheelslip. The TCU control unit restricted oil pressure in the hydraulic system so that although there was not enough pressure to lift the implement there was enough to transfer some of its weight onto the tractor. When wheelslip occurred the driver engaged TCU and adjusted a valve that varied the oil pressure in the circuit until wheel spin was hopefully eliminated. TCU could also be used with trailed implements by using a special three-point linkage drawbar hitch that partially lifted the front of the implement and transferred some of its weight on to the tractor.

Draft and position control hydraulic systems were standard on all new wheeled tractors in the early 1960s. Many had live hydraulics provided by an engine-mounted pump or a dual clutch. Three-point linkage for crawlers was still in the development stage but Ernest Doe & Sons and Simba were two companies which partly solved the problem with a trailed tool carrier on pneumatic wheels for use with tracklayers, the Doe Triple D and other big four-wheel drive tractors. The tool carrier had a category II three-point linkage

operated by hydraulic rams connected to the tractor's external services coupling.

When Warwickshire farmer Roger Dowdeswell was unable to buy a three-point linkage unit for his tracklayer he built one in his farm workshop. The manufacturing rights were sold to Turner Engineering at Alcester in the late 1960s and fully mounted implements could at last be used on crawler tractors.

Quick hitches, which made it easier and safer to attach an implement to the three-point linkage, were on the market in the mid-1960s. The Canadian-designed Salopian Kenneth Hudson Insta-Hitch and the German Accord automatic coupler had triangular frames attached to the three-point linkage and the implement. The triangular frame hitching system was said to complete the job in a matter of seconds. After lowering the linkage careful reversing was needed to align the two frames before raising the lift arms to attach the implement and lock it in position.

Alternative quick-hitch systems used various designs of extending the lift arm end with hinged claws that clamped over a ball or bobbin fitted to the lower hitch pins on the implement. The John Deere quick coupler consisted of a frame with one upper and two lower hook-shaped jaws attached to the three-point linkage. The hooks were located on the implement hitch pins and secured with a spring-loaded locking mechanism. Another system developed by K-H-D in Germany for Deutz tractors had similar hook-shaped ends on the lower lift arms that engaged with bobbins on the implement. With the lower links in place, a telescopic mechanism simplified the task of connecting the top link. The Perry-Jeffes hitch sold by Farm Implements Ltd at Stroud in Gloucestershire used extendable hooks on the lower links that engaged with balls fitted to the implement lower link pins. A hook on the implement end of the top link was located on the headstock pin and locked in position with the aid of a rope.

Massey Ferguson's early 1960s Pressure Control system is one of the

141. The Teagle quick hitch 'A' frame.

better-known devices for transferring some of the weight of a trailed implement onto the tractor to reduce wheelslip but there were others. The Gale Accord Tracassistor invented by Wiltshire farmer WA Gale and the Salopian Kenneth Hudson Insta-Hitch both achieved the same result. The Tracassistor had four heavy coil springs, connected from the top link position on the Alpha Accord quick-hitch frame to a hook welded on the implement drawbar. Weight was transferred onto the tractor when the springs were extended by slightly raising the raising lower lift arms. The Insta-Hitch, with a special hitch from the top link pin on the frame to the drawbar or pick-up ring hitch transferred some of the weight of a trailed implement on to the tractor. More advanced hydraulic systems with lower link sensing appeared in the 1970s and by the early 1980s tractor drivers had the facility to use a mix of draft and position control. Electronic draft control systems became standard and the linkage could be raised or lowered with remote control switches on the rear mudguard when hitching heavy mounted implements. The ever-increasing size and weight of mounted implements resulted in the almost universal use of automatic ball and claw implement hitches on most tractors by the late 1980s.

IMT

The Metalski Zavodi (factory) established in Belgrade in 1949 became Industrija Masina Tractora (Industrial Motor Tractors) in 1954. The Massey Ferguson connection with IMT began in 1955 when the grey and gold Ferguson FE35 superseded the TE20 series and an agreement with IMT resulted in Massey Ferguson manufacturing the Ferguson TE20 tractor, badged as the MF20, in Belgrade from 1956.

Approximately 2,000 MF20 tractors, some with Perkins diesel engines, were made in the first year of production. This arrangement also enabled Massey Ferguson to sell tractors in Eastern Europe where restrictive trade tariffs prevented normal export trading.

The IMT factory was eventually taken over by the Yugoslavian state and in a joint venture with Massey Ferguson the MF35, badged as the IMT539, was also built in Belgrade. Production continued under licence and apart from the use of metric threads the IMT539 was very similar to the MF35 but without the option of Multi-Power transmission. Several thousand IMT539 tractors had been sold when Vowcrest at Stone in

142. The Massey Ferguson 35 was made under licence in Yugoslavia as the IMT 539.

143. Front-wheel drive was automatically engaged when the IMT577DV had wheelslip.

Staffordshire introduced it to UK farmers in 1975. The tractor was popular with livestock farmers and an advertisement in the late 1970s suggested that 'although a dog is a livestock farmer's best friend, the introduction of the IMT539 has put the dog's position in danger'. With 39 hp under the bonnet, a dual clutch, ground speed power take-off and a differential lock it was 'a pretty versatile little character', and having been developed over a period of thirty years, the 539's traditional design was 'just as reliable as a dog and cost less to keep'!

The 50 hp IMT555 was introduced in 1963 and other models based on Massey Ferguson tractors appeared in the mid-1960s. The agreement with Massey Ferguson came to an end in 1968 but their influence could be seen in new IMT tractors launched in the 1970s. The IMT name was well established in the UK by 1980 when Vowcrest were importing eight models between 39 and 220 hp. They included the new 47 hp IMT549 and the articulated four- or eight-wheel drive IMT5200 based on the MF1500.

Vowcrest changed their name to IMT Tractors in 1982 when seven models including the two-wheel drive 39 hp 539, the four-wheel drive 110 hp 5106 and the four-wheel drive 5200 with a 207 hp diesel engine, which cost £38,500, were included in the price list. Most IMT tractors were shipped to the UK in a semi-finished state and the importer added a Perkins or Mercedes engine with Lucas or CAV electrical and fuel injection equipment. Additional soundproofing was installed around the IMT539 engine to reduce the noise levels so that the tractor conformed to noise regulations and could be sold in Britain with a fold-down safety frame instead of a quiet cab. New models introduced to the UK market included the 77 hp IMT577 and the four-wheel drive 577DV with ten forward and two reverse gears shown for the first time at the 1982 Royal Smithfield Show. Optional extras for the 111 hp IMT5106 and 138 hp 5130 included a partly synchronised ten forward and two reverse gearbox and a twenty-four speed synchronised transmission with a reverser.

A modified range of 37–70 hp economy tractors with

new cabs was introduced in 1986 when the IMT549 and 569 were improved with new direct injection engines and hydrostatic steering. An advertisement suggested that by buying an economy range IMT tractor the days of the more expensive tractor would be numbered and it was time for farmers to 'start ploughing money back where it was needed most – in their pockets'.

IMT Tractors moved to Scunthorpe in Lincolnshire in 1988 when the 539's steel bonnet and mudguards were replaced with fibreglass. The IMT585 was the top model in

144. An air compressor for trailer brakes and inflating tyres was an optional extra for the IMT549.

the 59–85 hp two- and four-wheel drive 5 series announced in 1989. It had an 85 hp turbocharged direct injection diesel engine, ten forward and two reverse gears and a heated cab. The basic cost of the two-wheel drive 585 was £9,999 and in 1989 a free pasture topper or post-hole digger was included in the price. IMT Southern at Sturminster Newton in Dorset and IMT Northern at Nantwich in Cheshire were importing the Belgrade-built tractors when the 100 hp six-cylinder IMT5106 was introduced in 1991. The four-wheel drive tractor had a ten forward and two reverse mechanical gearbox, hydrostatic steering, wet disc brakes, a two-speed power take-off and two hydraulic pumps. One supplied oil to the electronically controlled hydraulic linkage and the second provided oil pressure for a bank of three external services spool valves. The IMT539 was still being made in 1991 with an improved specification which included hydrostatic steering with an optional belt pulley and engine-mounted compressor for air-operated trailer brakes. There were eight IMT tractors on the UK market in 1992 and with the exception of the 43 hp two-wheel drive 542, the others were all available with two- or four-wheel drive.

Production continued on a small scale during the war in Yugoslavia but trade sanctions imposed in 1993 prevented the export of tractors until 1998 when IMT Southern re-introduced the 539 with an optional safety cab to the UK market. The three-cylinder IMT539 with

a six forward and two reverse gearbox, power steering and full road lights cost £6,800. More than twenty IMT models were being made in the former Yugoslavia by the end of that year but the 539 was the only one available in the UK. The full range included fruit, high visibility and standard models up to 90 hp with optional four-wheel drive and a Perkins, Mercedes or IMR engine built in Yugoslavia under licence from Perkins. Transmission systems ranged from a six forward and two reverse mechanical gearbox to a twelve-speed synchronised box with reverse shuttle. Other alternatives included a semi-synchromesh ten forward and two reverse gearbox and a fully synchronised twenty-four speed box with reverse shuttle. Front power take-off and linkage, automatic implement hitching and electronic linkage control hydraulics were optional.

International Harvester

The McCormick Harvesting Machine Co, the Deering Harvester Co and three smaller American harvest machinery manufacturers merged in 1902 to form the International Harvester Company. They made their first farm tractor in 1906, which, like other internal combustion engined tractors of the time, looked more like a traction engine than a farm tractor. The 20 hp Mogul with an open chain drive to the rear wheels introduced in 1909 followed this trend but the 15-30 launched in 1921 and the 10-20, which followed in 1923,

145. The W-12 was one of the models in the range of 1930s International Harvester ploughing tractors.

had a more modern appearance. The 10-20, so called because it developed 10 hp at the drawbar and 20 hp at the belt pulley, had a four-cylinder paraffin engine with magneto ignition. About 6,000 International 10-20s were sold to British farmers between 1923 and 1939. About 300 imported after 1938 had pneumatic tyres while the others were on steel wheels. The 22½ hp McCormick-Deering Farmall Regular, so called because it was claimed to be the first tractor suitable for all types of farm work, was introduced in 1924.

Farmers wanted more powerful tractors in the late 1920s and the three-plough Farmall F-30 launched in 1931 and the F-20 added in 1932 met this demand. The smaller 12 drawbar hp F-12 with a four-cylinder petrol or paraffin engine on pneumatic tyres or steel wheels introduced in 1932 was one of the more successful rowcrop tractors of the period. The Farmall F-14 with a higher engine speed, which developed 14.8 drawbar hp, replaced the F-12 in 1938. The power take-off and belt pulley were standard but hydraulic linkage was an optional extra. International Harvester tractors were painted grey until 1936 and the F-12 was one of the first models with International red paintwork.

International Harvester also made a range of

ploughing tractors including the W-12, W-14, W-30, W-40 and the diesel-engined WD-40 in the 1930s. The W-12 and W-14 shared the same engine and a three forward speed and one reverse gearbox with the F-12 and F-14. The four-cylinder W-30, made between 1932 and 1940, developed 31 belt hp and 19 drawbar hp and the three forward gears and one reverse gearbox had a top speed of 3¾ mph. The six-cylinder W-40, built from 1935 to 1940, was rated at 46¾ belt hp and 35¼ hp at the drawbar. The WD-40 introduced in 1934 and claimed to be America's first diesel-engined wheeled tractor developed 48¾ belt and 37¼ drawbar hp when it was tested at Nebraska. The engine was started by hand on petrol and switched to diesel when it was warm. There was a carburettor and magneto on one side of the engine and the diesel injection pump, filters and injectors were on the opposite side. The engine had separate inlet valves for running on petrol and on diesel. A hand lever was used to reduce the compression ratio to 6:1 for starting and when the engine was warm the lever was used to isolate the inlet valve for running on petrol and increase the compression ratio to 14:1 for diesel operation.

McCormick-Deering tractor model letters were self-

146. The late 1930s T-6 was the vaporising oil version of the diesel-engined International TD-6 crawler tractor.

explanatory. W was for wheeled models, T for tracklayers and D denoted a diesel engine. The first McCormick-Deering crawler tractors were made in 1928. The 10–20 and 15–30 TracTracTors with the rear track sprockets larger than those at the front were based on the 10–20 and 15–30 wheeled models. The T-20 crawler introduced in 1931 with a Farmall F-20 engine had a more conventional track layout and the bigger T-40 and TD-40 crawlers based on the W-40 and WD-40 were made between 1933 and 1939. More agricultural and industrial crawlers, including the T-6/ TD-6, T-9/TD-9, T-14/TD-14 and the TD-18, were launched in America in 1938 and 1939. The TD-14A and TD-18A replaced the earlier models in 1949.

The diesel engines were started on petrol but unlike earlier International Harvester diesel engines they were automatically switched over to diesel after they had run for about 900 revolutions. The TD-6 and TD-9 specification included a five forward and two reverse gearbox, clutch and brake steering, a power take-off and optional belt pulley. The TD-9, made between 1939 and 1956, had the same four-cylinder overhead valve engine

as the WD-9 wheeled tractor. The 85 hp TD 18, which weighed just over 10 tons and used more than 6 gallons of fuel per hour, was more than three times heavier than the TD6 with a full tank of diesel fuel.

The Farmall F series was superseded by the Farmall A, B, H and M tractors which were introduced to American farmers in 1939. The A and B were advertised as 'one-plow tractors', the 26 hp Farmall H was a 'two-plow model' and the 'three-plow' Farmall M had a 36 hp engine. The Farmall A and B had the same 18 hp petrol or optional vaporising oil engine but the A had a 40–68 in wheel track adjustment compared with 64–92 in for the Farmall B. The seat and driving controls were offset and this design, called 'Cultivision', gave the driver an unobstructed view when working in rowcrops. The Farmall A and B were available with a tricycle or standard four-wheel front axle and the high clearance Farmall AV had larger diameter wheels with extended front axle stubs. The Super A with four forward gears, a top speed of 10 mph, electric lights, electric starting and Touch Control hydraulics replaced the Farmall A in 1947. The Farmall C and Super C, made between 1948

147. The 'Cultivision' Farmall A with an offset seat gave the driver an unobstructed view when working in rowcrops.

148. The Farmall H was introduced to American farmers in 1939.

149. The 9 hp Farmall Cub, suitable for farms of up to 40 acres, had an offset driving seat making it ideal for rowcrop work.

and 1954, also had the same engine as the Farmall A and B with a similar chassis to the Farmall H and M and the rear wheel track was adjusted by sliding wheel hubs in or out on their axles. The Farmall Cub, the smallest American Farmall and made between 1947 and 1954, had a 9 hp four-cylinder engine, a three forward and one reverse gearbox with a top speed of 6½ mph and a 40–56 in wheel track. Touch Control hydraulics, a power take-off and a belt pulley for mounting on the power shaft were available at extra cost. More powerful Farmall Cub tractors were made in the late 1950s and by the 1970s a lightweight Farmall Cub with increased engine power was still available in America.

The Farmall M with a five forward and one reverse gearbox was one of the most popular International tractors in Britain and America. Many Farmall M tractors made during World War Two had steel wheels with the fifth gear blanked off to reduce the top speed to under 6 mph. Most Farmall H and M tractors, including the high clearance HV and MV rowcrop models with tricycle front wheels, had pneumatic tyres in the mid-1940s. Optional equipment included power take-off, belt pulley, the Lift-All hydraulic system, electric starter and

lights. The diesel-engined Farmall MD launched in 1941 was replaced by the Super MD in 1952 and this tractor was made until 1954.

The high ground clearance Farmall H and M were well suited to rowcrop work and the McCormick Deering W-4, W-6 and W-9 made between 1940 and 1953 were ideal ploughing tractors. The W-4 had the same vaporising oil engine as the Farmall H and the Farmall M and MD spark ignition engines were used for the W-6 and WD-6. The W-9 and the WD-9 were rated at 52 hp and the diesel-engined WD-9, like the earlier TD6 and Farmall MD tractors, was started on petrol and automatically switched to diesel when the engine was warm. The W-4 and W-6 had five forward gears and a top speed of 14 mph, the W-9 was 1 mph faster in top gear and like the Farmall the fifth gear was blanked off on tractors with steel wheels. Specialist versions of the W series tractors were also made. The industrial ID-4, ID-6 and ID-9 models had heavy cast-iron wheel centres and modifications for the O-4 and O-6 orchard tractors included a down-swept exhaust, relocated air cleaner, guards over the rear wheels to prevent tree damage and a shield over the steering wheel. The cheaper OS-4 was

150. The W-4 had a top speed of 14 mph but the top gear was blanked off on tractors with steel wheels.

similar but without tree guards for the rear wheels and steering wheel.

Construction of a new International Harvester factory at Doncaster began in 1939 but with the onset of World War Two it was used to manufacture munitions. The first International B-8A ploughs were made at the Wheatley Hall factory late in 1948 followed by the B-R green crop loader early in 1949. Tractor production was delayed until September 1949 when the Minister for Agriculture drove the first British-built Farmall M off the Doncaster assembly line. The 37 hp vaporising oil-engined tractor with a five forward and one reverse gearbox, power take-off and side-mounted belt pulley became the Farmall BM in 1951 to denote it was made in Britain. The Farmall M and H were still being made in America when the BMD with a 38 hp British-built diesel engine appeared in 1952 and with further improvements in 1953 the tractors were re-badged as the Super BM and Super BMD.

There was choice of a 50½ hp petrol or a 42½ hp vaporising oil engine for the Super BM and the 50½ hp Super BMD with an indirect injection diesel engine had a glow plug (heater) in each cylinder for cold starting. A maximum 88 in rear wheel track setting was achieved by sliding the wheels in or out on their axles and by reversing the centre discs. The front track width was adjustable in 4 in steps from 51 to 81 in and live hydraulics with an engine-mounted pump were optional for Super Farmall tractors. A constant supply of oil was pumped from a separate reservoir through a control valve to an external double-acting ram, which raised and lowered the lift arms. The ram was also used to alter the height of the drawbar, a useful aid when hitching trailed equipment.

The Doncaster factory was in full production by 1954 when the new Farmall 100, 200, 300 and 400 tractors with 20–50 hp engines were launched in America. Most new tractors sold in Britain had a diesel engine when the Doncaster-built 50½ hp Super BWD-6 was launched in 1954. The Super BW-6 with a vaporising oil engine was added in 1955 but very few were sold. The new tractors had Super BM and Super BMD engines, five forward gears and one reverse, live Lift-All hydraulics, a 540 rpm power take-off and a belt pulley. The BW-6 and BWD-6 were ploughing tractors and the heavy cast-iron rear wheel centres limited track width adjustment to 56, 60

151. A glow plug heater in each of the Farmall BMD cylinders was used to help start the diesel engine when it was cold.

and 64 in. Power take-off speeds were at last falling into line with international standards and Super BWD-6 sales literature pointed out that the power take-off with safety shield conformed with current SAE and BSI standard speeds. The last Super BWD-6 and Super BW-6 tractors were made in 1958.

More than three million International Harvester tractors had been produced worldwide when the B250, the first British-built medium-powered International Harvester tractor, was launched at the 1956 Royal Smithfield Show. Made at the old Jowett car factory in Bradford, the B250 had a 30 hp four-cylinder indirect injection diesel engine with glow plug cold starting, a five forward and one reverse gearbox, live dual category hydraulics with an engine-mounted pump and automatic pick-up hitch. It was one of the first wheeled tractors to have a diff-lock and its disc brakes were claimed to be equally positive in forward or reverse. Standard and high-clearance versions of the more powerful B275 with a 35 hp diesel engine and an eight forward and two reverse gearbox were added in 1958. The International B250 remained in production until 1961 and the B275 until 1968.

The Farmall B450, built between 1958 and 1970, was a re-styled version of the BWD-6 with a four-cylinder 55 hp indirect injection engine with a pneumatic governor, a five forward and one reverse gearbox diff-lock and self-energising disc brakes. An advertisement explained that 'the thin oil, which gives the hydraulic system an instant all-weather response is stored in a separate reservoir and supplied by an engine-mounted pump to the live category II hydraulic system'. The B450 had a new two-stage hydraulic response control designed to improve traction in tough conditions. The first two-thirds of control lever movement progressively transferred implement weight on to the tractor and the final third raised it from work.

Power take-off and dual category live hydraulics linkage were still optional for International Harvester tractors in 1959. The B250 cost £510 and the B275 £450, power take-off and hydraulic linkage added £57 and £78 respectively and although it appears rather mean an extra £2.13s was added to the bill for an hour meter. The Farmall B450 cost £795 plus £90 for the power take-off and hydraulics but at least the hour meter was included in the price!

152. Launched in 1958 the Farmall B450 had a four-cylinder indirect diesel engine, five forward gears and one in reverse.

The British-built BTD6 and BT6 crawlers launched at the 1953 Royal Show in Blackpool were similar to the American TD6, which had been available in the UK for several years. The BTD6 had a modified 38 hp Farmall BMD indirect injection engine with electric starting, five forward gears and one reverse, clutch and brake steering and an hour meter. The Farmall BM vaporising oil engine was used for the BT6 crawler. Improved versions of the BTD6 with a 50 hp engine, a strengthened transmission and a spiral bevel crown wheel and pinion in place of the earlier straight toothed gears appeared in 1955. The 40 hp BTD640, a basic version of the BTD6 with lighter tracks introduced in 1957 for farm work, cost £1,500 but the price didn't include an exhaust silencer. Intended mainly for drainage work and the construction industry the BTD20 crawler with a 124 hp Rolls Royce engine, launched in 1958, had a six forward and six reverse speed transmission. Designed for agricultural work the 40 hp BTD5, also based on the BTD6, was added to the International crawler range in 1964. An optional offset frame for the hydraulic linkage made it possible to keep both tracks on level ground when ploughing.

Sales literature described the McCormick International B414 introduced in 1961 as 'the tractor with everything you've wanted most'. Farmers were consulted, a report was written and International Harvester then 'went ahead and actually built the

tractor'. Features that farmers suggested and got on the B414 included a 36 hp diesel engine, eight forward and two reverse gears, diff-lock, Vary-Touch automatic draft control hydraulics, self-energising disc brakes and a more comfortable seat. There was also a petrol-engined B414 and a narrow model for orchard work. The 60 hp B614 launched in 1963 was similar to the earlier B450 with live hydraulics and an independent 540 and 1,000 rpm power take-off with multi-plate clutch. Two power take-off stub shafts with six and twenty-one splines were supplied with the tractor and the required shaft was bolted on to a hub at the back of the tractor. The 56 hp 634, which replaced the B614 and was made between 1968 and 1972, had an indirect injection engine, eight forward and two reverse gears, Vary-Touch hydraulics and optional four-wheel drive.

The International 706 and 806 with 89 hp and 110 hp six-cylinder engines, launched in America in 1963, were introduced to British farmers a year or two later. Petrol- and diesel-engined versions were made and they were among the first tractors with hydrostatic steering. The 706 had an eight forward and four reverse gearbox, category II three-point linkage and an independent 540/1,000 rpm power take-off. The 806 had a torque amplifier giving sixteen forward speeds, twin disc hydraulic brakes and a torsion bar lower link sensing system was used for the hydraulic draft control. The 806 had three hydraulic pumps, one for the three-point

153. The three-cylinder International Harvester 523 was made in Germany.

linkage, another for the hydraulic power take-off clutch and the third serviced the hydrostatic steering and brakes. It was also one of the first tractors to have a dry element air cleaner with an automatic dust unloader. Both tractors were discontinued in 1967. Standard, high-clearance and narrow versions of the B434 superseded the B414 in 1966. Improved styling, better access for servicing and another 4 hp under the bonnet were the more obvious changes to the petrol- and diesel-engined B434. The German-built International Harvester 523 with a three-cylinder 52 SAE hp cost £3,400 when it was launched in 1965. An eight forward and two reverse gearbox was standard and the optional Agrio-matic-S transmission with a clutchless change in each gear doubled the number of speeds.

The National Institute of Agricultural Engineering at Silsoe developed an experimental hydrostatic transmission in the early 1960s and the International Harvester 656, introduced to American farmers in 1967, was the first production model with hydrostatic transmission. The 656 had an engine-driven hydraulic motor which transmitted power through a mechanical final drive system to the rear wheels. There were two speed ranges and a control lever provided maximum stepless forward top speeds of 8 and 20 mph. The reverse speeds were limited to 4 and 9 mph and a Foot-N-Inch pedal was used as an emergency clutch and provided precise control in confined spaces.

More new International tractors including the 354 and 374 with 35 hp engines and the 42 hp 444 appeared in 1970. The 'no frills' 374 had an eight forward and two reverse gearbox, live category I and II hydraulics and a two-door cab. The 354 and 444 were similar with an optional independent power take-off. The 454, 474, 574 and 674 with 50–71 hp engines were added to the range in 1972. An optional hydrostatic transmission controlled with a single lever on the dashboard was also available for the 474 and 574.

High horsepower two-wheel drive tractors were popular in America in the 1970s and the six-cylinder 106 hp International Harvester 966 Farmall launched there in 1971 was typical. The 966, which arrived in Britain in 1972, had an eight forward and four reverse gearbox, hydrostatic steering, hydraulic brakes, and optional hydrostatic transmission. Like the earlier 806 it had three hydraulic pumps for the power steering and brakes, a power take-off clutch and the category II lower link sensing hydraulic system. The rear wheels could be moved along on their axles to give track widths from 73 in to 102 in and the cab, built around a safety frame, could be tilted backwards for servicing the transmission. Many features of the 966 were retained when the 946 and 1046 replaced it in 1975 but the new models had six-cylinder 90 and 100 hp engines and a twelve forward and five reverse synchromesh gearbox.

International Harvester exhibited a nine-model tractor range, including the new Worldwide 84 Series Hi-Performers, at the 1977 Royal Smithfield Show. The two-wheel drive 35 hp 374, which cost £4,100, was the smallest and the four-wheel drive 136 gross hp 1246 with a £16,400 price tag was the most powerful and expensive model in the range. Five other 84 series tractors, together with the 955 and 1055, which

154. The International 784 was launched at the 1977 Royal Smithfield Show.

replaced the 946 and 1046, completed the list. The 374 and 384 were very basic two-wheel drive tractors and their standard specifications did not include a synchromesh gearbox, hydrostatic steering or an independent dual speed power take-off. The 77 hp Hydro 84 had an infinitely variable hydrostatic transmission, the other 84 series tractors had an eight forward and four reverse synchromesh gearbox with an optional torque amplifier to double the number of speeds. A twelve forward and four reverse gearbox was standard on the 1246. The International 1455, launched at the Royal Smithfield Show in 1981, had a 147 hp turbocharged six-cylinder engine and a sixteen forward and seven reverse speed gearbox but four-wheel drive and hydrostatic steering were not included in the £23,000 price tag. It had a fluid turbo clutch that enabled the drive to be taken up by simply increasing engine speed after selecting the required gear.

Five 85 series Field Force tractors, ranging from the 52 hp 485L to the 82 hp 885XL made at Doncaster between 1981 and 1985, were improved versions of the 84 series. Publicity material explained that other tractors had a cab but the 85 series was different with a 'new XL control centre (cab) with glass from the floor to the roof which had cost £28m to develop'. A full synchromesh gearbox with eight forward and four reverse speeds was standard and the optional change-on-the-move torque amplifier doubled the gear range. Four-wheel drive was available

for the 69, 77 and 82 hp 85 series tractors including the 77 hp Hydro 85 with hydrostatic transmission which replaced the Hydro 84. The 45 hp 385L was added to the 85 series in 1984. International Harvester marketed a range of 90–170 hp German-built tractors in the UK in 1981. The two- and four-wheel drive, six-cylinder 55 series launched in 1977 were improved with new XL cabs. The 955XL, 1055XL and 1255XL had 90, 100 and 125 hp power units and a 145 hp turbocharged engine was used for the flagship 1455XL. Turbo clutches, twenty forward and nine reverse gears and a 'buddy seat' were standard on the 1255XL and 1455XL.

The three articulated four-wheel drive International 'Snoopy tractors' launched in America in 1979 arrived in the UK in 1981. Classed as rowcrop tractors in America the six-cylinder 3388, 3588 and 3788 tractors rated at 130, 150 and 170 hp had a rear-mounted cab and a torque amplifier which doubled the number of gears in the eight forward and four reverse mechanical gearbox. The American-built International 5288 180 hp four-wheel drive tractor first seen at the 1982 Royal Smithfield Show was unusual in that instead of having the clutch mounted on the flywheel it was located on the gearbox layshaft. It had a range gearbox with three forward and one reverse synchromesh gears and with six sequential speeds in each range the 5288 had a total of eighteen forward and six reverse speeds.

155. The 52 hp 885 was made at Doncaster between 1981 and 1989.

156. The German-built 1455XL, the flagship model in the International Harvester 85 series, had a 145 hp six-cylinder engine.

157. Power steering and oil-immersed multi-disc brakes were features of the Iseki SX65.

Tenneco, who already owned Case and David Brown, increased their product range in 1984 when they acquired International Harvester's agricultural division. Manufacture of the re-badged Case IH85 series continued at Doncaster, the David Brown Case 94 series was still made at Meltham and the big Case 4694 and 4894 tractors with 250 and 291 hp engines were built in America. The 85 series remained in production until 1989 when the Case IH Magnum and Maxxum series tractors replaced them.

Case IH and New Holland merged in 1999 to form CNH Global and a condition of the merger required CNH Global to sell the Doncaster factory along with the production rights for the Case-IH CX and MC-X series tractors. The Italian ARGO Group, already owners of Landini and Valpadana, formed McCormick Tractors International Ltd at Doncaster when they bought the factory and tractor production rights in 2001.

Iseki

The Iseki Agricultural Machinery Manufacturing Co was established in Japan in 1926 but Iseki tractors were not sold in Great Britain until the mid-1970s. Tractors had become more powerful over the years and by this time most of them were not really suitable for small farms and this provided an opportunity for Iseki, Kubota and other Japanese tractor manufacturers to sell compact tractors in the 10–20 hp bracket in the UK.

Following a visit to Japan in 1975 Len Tuckwell, a John Deere dealer in Suffolk, started a company called LE Toshi Ltd at Ipswich to import the two- and four-wheel drive 13 and 15 hp Iseki TS and TX tractors with Mitsubishi two-cylinder water-cooled engines. The TS2810 and TS3510 with Isuzu water-cooled engines were soon added to the UK range. An eight forward and two reverse gearbox, live hydraulics, a four-speed power take-off and a safety start mechanism in the clutch pedal linkage were features of the 28 and 35 hp Iseki tractors.

Iseki were still making the same tractors when they established a partnership with Lely at St Neots in 1979. Within a year an extended range included the two- and four-wheel drive 13 hp TX1300 and TX1300F, the 15 hp four-wheel drive TX1500F and five 19–35 hp Iseki TS models. The smallest TS1910F had a 19 hp engine, the new two-wheel drive twin-cylinder 22 hp TS2200 was next in line and the two-wheel drive TS3510 was the

158. The JCB High Mobility Vehicle made its debut at the 1990 Royal Smithfield Show. It was capable of towing 14-ton loads on the road at 40 mph.

most powerful tractor in the Lely Iseki price list. The 35 hp TS3510 with a dual clutch, live hydraulics and a quiet cab cost £5,800. New Iseki tractors with bigger engines, still classed as compact tractors, appeared in Britain in 1982. The 1.5 m wide two- and four-wheel drive TS4510 had a 45 hp three-cylinder water-cooled engine with a special combustion chamber designed to improve fuel economy. The eight forward and two reverse gearbox gave a top speed of 13 mph and the TS4510 could be supplied with a rollbar or a quiet cab. The 65 hp four-wheel drive Iseki TX6500 had a similar specification to the TX4510 and like the high-clearance version with four equal-sized wheels it had twenty forward gears and the choice of standard, rowcrop or low ground pressure wheels.

The Lely Iseki arrangement continued until 1986 when Iseki established an import business near Huntingdon and later at Bourn in Cambridgeshire. Iseki entered the big-power league in 1986 with four new SX models with 50–95 hp engines. The SX tractors had a twelve forward and four reverse or a twenty-four forward and eight reverse gearbox with Super-Shift transmission. Other Iseki models on the British market at the time included the three-cylinder 16 hp TX2140 and the 18 hp TX2160 together with five TE models in the 24–48 hp bracket. The TX2160 had closed centre hydraulics and optional hydrostatic transmission and the

four-wheel drive TA series tractors with 25, 33, 40 and 48 hp engines introduced in 1989 had a sixteen forward and reverse speed mechanical gearbox and power-assisted steering.

Massey Ferguson signed an agreement with Iseki in 1993 to provide the 88–125 hp MF3000 and MF3100 tractors with crystal blue paintwork for sale in Japan and in return Iseki supplied 25–48 hp TA tractors with the MF colour scheme for the UK market. This arrangement ended in 1997 when Iseki TF and TM tractors were added to the Jacobsen range of compact tractors. The TF tractors had 20, 25 and 30 hp engines and there was a choice of a mechanical or hydrostatic transmission for the 15 and 17 hp TM models with three-cylinder water-cooled engines. Jacobsen was part of the American Textron Group and when Ransomes was acquired by Textron in 1998 Iseki compact tractors were distributed by Ransomes Jacobsen at Ipswich.

JCB

Bamfords of Uttoxeter were a leading farm machinery manufacturer in the early 1900s but Joe Bamford left the family business in 1945 to establish his own agricultural engineering company. Early products included tipping trailers and front-end loaders for the E27N Fordson Major and Nuffield tractors. Half-track conversions with slatted rubber tracks and the single arm Si-Draulic front-end loader came next. By the mid-1950s the Bamford product range had been widened to include loader shovels and the first JCB digger-loaders.

Field trials of a prototype JCB High Mobility Vehicle began in 1987 and the pre-production HMV or Fastrac unveiled at the 1990 Smithfield Show was probably one of the worst-kept secrets in the history of farm tractors. The initial image of the Fastrac, which eventually went on sale in the summer of 1991, was that of a fast transport tractor rather one than for ploughing and heavy field work. Innovations on the Fastrac, advertised as 'the world's first genuine high-speed tractor', included coil spring front suspension, self-levelling hydro-pneumatic rear suspension, air-operated disc brakes on the four equal-sized wheels and an air-conditioned two-seater cab. The tractor could have a naturally aspirated 125 hp or a turbocharged 145 hp Perkins diesel engine. Other features included an eighteen forward and six reverse gearbox, air brakes on all four wheels, front and rear power take-off shafts and hydraulic linkage with the Bosch Hitch-Tronic draft control system.

The improved Fastrac 125.65 and 145T.65 were announced by JCB Landpower Ltd at the 1992 Royal Smithfield Show. Different wheel equipment and gearing gave the tractor a top speed of 40 mph, the hydraulic linkage was redesigned to take category III mounted implements and there was also a choice of a light- or heavy-duty rear load platform. There were three models of the JCB Fastrac in 1993 when the 125 was joined by the new 135 hp Fastrac 135 and the 150 hp Fastrac 155 which replaced the 145T.65. An improved transmission had a thirty-six forward and twelve reverse Powersplit change-on-the-move gearbox and an optional pre-select Selectronic system with a column-mounted gear lever, which provided rapid changes of speed range and direction. An even more powerful Fastrac 185 with a six-cylinder turbocharged 170 hp Cummins engine made its debut at the 1994 Royal Show. The new model had an improved transmission and rear axle and there was a choice of a 65 or 75 kph gearbox. The diff-lock was automatically disengaged when the hydraulic linkage was raised and the optional field area meter stopped measuring at the same time.

The smaller and lighter Fastrac 1115 with a 115 hp Perkins engine replaced the Fastrac 125 in 1995 and the similar 135 hp 1135 was added to the compact range later in the year. The larger 130, 150 and 170 hp Fastracs had four-wheel hydraulic power braking and twin-line air braking system for trailers.

159. The mid-1990s Fastrac 185-65 with a 170 hp Cummins engine had 36 forward and 12 reverse gears. The '65' in the model number indicated that the tractor had a top speed of 65 kph.

160. Two-wheel, four-wheel and crab steering was a feature of the 125 hp JCB 1125.

Changes in 1996 included a new engine for the 1115, which became the 1115S, the addition of the 125 HP Fastrac 1125 and the existing 1135 completed the 1000 series. They were all available with two- or four-wheel drive and a new air-over-hydraulic braking system replaced the previous hydraulic power brakes. A Quadtronic two-wheel, four-wheel and crab steer option, also introduced in 1996, gave JCB their first fully suspended all-wheel-steer fast tractor with a much reduced turning circle.

The Fastrac was well established in 1997 with six models having six cylinder engines in the 115–170 hp bracket. The 1000 series met the demand for a smaller high-speed tractor and the 100 series provided more power for heavier work. The 100 series consisted of the 135 and 155 with Perkins 135 and 150 hp engines and the 185 had a 170 hp Cummins turbocharged engine. With the weight evenly distributed over the four wheels, a full suspension system and air-operated disc brakes made the tractors legal for use on the road at speeds of up to 50 kph (31 mph).

The Fastrac 100 and 1100 models were improved, re-styled and re-badged in 1998 as the 2000 and 3000 series with a fifty-four forward and eighteen reverse powershift transmission. The 155 and the 185 became the 3155 and the 3185 and four 2000 series tractors in the 115–148 hp bracket included the new Fastrac 2150. With the exception of the 3185 with a 170 hp Cummins engine, the JCB 2000 and 3000 series had turbocharged Perkins power units. The 148 hp Fastrac 2150 was added to the range to meet the needs of existing 1135 owners who wanted the power of the 3155 with the compactness and manoeuvrability of the previous 1000 series tractors.

Six JCB Fastracs with turbocharged engines were current in 2000. The 3155 had a 150 hp six-cylinder Perkins engine and a 170 hp Cummins provided the power for the 3185. The four 2000 series Fastracs had 115–148 hp Perkins engines and all of them had a fifty-four forward and eighteen reverse powershift transmission with twenty-one of the gears in a 2½–7½ mph range for fieldwork. JCB Fastracs in 2004 included the 2140 with optional Quadtronic steering together with the 3170, 3190 and 3220 with 155–220 hp engines and a fifty-four x eighteen powershift transmission. Launched in 2005, the Fastrac 8250 with a Tier 3-compliant 248 hp Cummins six-cylinder common rail diesel engine and CVT transmission and capable of speeds up to 50 mph had a full suspension system, ABS disc brakes and touch screen controls in the cab.

John Deere

John Deere, who started a blacksmith's business in Grand Detour, Illinois in 1836, made his first steel plough in 1837 and he was making cultivators and corn planters in 1859 when he moved to Moline in Illinois. John Deere showed some interest in making tractors in 1892 but little happened in this field until the company bought the Waterloo Gasoline Tractor Co at Waterloo in Iowa in 1918.

The Waterloo Boy Model N-R tractors were sold in the UK as the Overtime N and R by Associated Manufacturers. The Model N was superseeded in 1923 by the 15–27 hp John Deere Model D. It was the first of many John Deere tractors with a horizontal twin-cylinder engine and an offset power shaft. An improved 24–37 hp Model D with a three-speed gearbox appeared in 1935

and with further improvements in 1939 the 30–38 hp Model D was made until 1953. The 19–25 hp John Deere Model C rowcrop tractor, launched in 1927, was re-badged as the John Deere GP in 1928.

Several versions of the tricycle-wheeled 18–24 hp John Deere Model A introduced in 1934 included rowcrop, wide, narrow, high-clearance and orchard models and some were made until 1953. The Model A had pneumatic tyres and the rear wheels could be moved in or out on their axles to adjust the track width. The tractor had a mechanical implement lift mechanism but unlike earlier models the power take-off was in a central position between the rear wheels. The smaller 12–16 hp Model B made between 1935 and 1952 was advertised as a tractor 'able to do as much work as two horses'. Styled versions of the Model A and Model B with a new four-speed gearbox were introduced in 1938 and a two-range six forward speed gearbox was added in 1941. So-called styled tractors had a bonnet and radiator grille to give them a more modern look compared with earlier models where the engine, radiator and steering mechanism were exposed to the weather.

The more powerful 'three plow' 31–41 hp John Deere Model D was the biggest tractor in the range in 1935. To

161. The John Deere-Lindeman BO crawler, for orchard work, was a tracked version of the Model D.

162. The offset seat on the John Deere Model L provided the driver with a good view when working in rowcrops.

meet a demand from American farmers for less powerful tractors the AR and BR were introduced in 1934. The 'two-plow' AR and BR, with steel wheels or pneumatic tyres, were general-purpose versions of the Model A and B. The BR was discontinued in 1947 but the AR was made until 1953. Other versions of the Model A and B included the narrow rowcrop AN and BN, the wide rowcrop AW and BW and the AV and BV for vineyard work. The AO and BO orchard tractors had downswept exhausts and independent foot brakes. Lindeman Power Farm Equipment, who converted the John Deere Model D tractors to tracklayers in the early 1930s, also fitted tracks to the John Deere Model B tractor. The 14–18 hp John Deere-Lindeman BO crawler made from 1939 until 1947 had a twin-cylinder paraffin engine, four forward gears and one reverse, clutch and brake steering and adjustable track centres.

The John Deere Model L, introduced in 1937, was advertised as a replacement for the horse on small American farms. It had a vertical twin-cylinder 9–10½ hp Hercules engine and a three forward and one reverse gearbox. The seat was offset to the left and the steering wheel offset to the right to provide good visibility when working in rowcrops. A 10 hp John Deere power unit replaced the Model L's Hercules engine in 1942. The 35

hp Model G rowcrop tractor with four forward gears and one reverse appeared in 1938 and the 13–15 hp Model H rowcrop tractor launched in 1939 had the same gearbox and a very modern foot throttle. The 12 hp LA rowcrop model, also with an offset driving seat, was added in 1941.

Horizontal twin-cylinder petrol engines were still used in 1947 when John Deere introduced the 20 hp Model M with a power take-off, electric starting and lights and the new Touch-o-Matic hydraulic system for three-point linkage mounted John Deere implements and to raise and lower the drawbar. The Model M, the MT tricycle-wheeled rowcrop tractor and the MC crawler introduced in 1949 were made for the next three years. Jack Olding & Co at Hatfield advertised some good news for farmers in 1947 when they announced that John Deere Model A and Model D tractors along with the 12A combine harvester and two models of John Deere plough would be available in 1948.

The 48 hp twin-cylinder Model R, introduced in 1949, was the first diesel-engined John Deere tractor. It was started with a donkey engine (known as a 'pony' engine in America) and had a five forward and one reverse gearbox, power take-off and improved Power-trol hydraulics controlled from an 'armchair driving seat'. An independent power take-off and a live hydraulic system were optional for the Model R which was the last of the letter series of John Deere tractors. Model numbers were used when the 31 and 41 hp John Deere 50 and 60 replaced the Model A and Model B in 1952. Rear wheel track widths on the Model 50 were adjusted with a ratchet mechanism that moved the wheels along the axles on a very coarse thread.

The standard, rowcrop and high-clearance versions of the 39 hp Model G became the Model 70 in 1953 when the Model M was rebadged as the John Deere

163. The twin-cylinder Model R was the first John Deere diesel-engined tractor.

Model 40. An improved Model 70 with a 43 hp vaporising oil, 50 hp petrol or diesel engine was added in 1954. The diesel tractor also started with a pony engine had a decompressor to help the pony engine get the main engine up to the required starting speed. The twin-cylinder 67 hp Model 80 diesel tractor with a pony engine, which superseded the Model R in 1954, had a six forward and one reverse gearbox and live hydraulics.

John Deere introduced the twin-cylinder 20 series tractors in 1956. The 25 hp Model 40 re-badged as the John Deere 320 was the smallest and the new 29 hp five forward speed 420 available in standard, rowcrop, tricycle rowcrop, high-clear and tracklayer format was next in line. The petrol-engined John Deere 520, 620 and 720 were rated at 38, 48 and 59 belt hp. Slightly more powerful liquefied petroleum gas (LPG) engines were available but LPG was not available in all parts of America. Power steering was standard on the 520, 620 and 720. The 67 hp diesel-engined Model 80, also with power steering, became the John Deere 820 in 1956 and an engine change in the following year boosted the power output to 75 hp. Standard features on the 20 series included draft control hydraulics, a pedal to engage drive to the power take-off, a more comfortable driver's seat and a new colour scheme with yellow flashes added to the familiar green paintwork. The restyled 20

series with the same engines, live 540/1,000 rpm power take-off and improved driving controls and operator comfort, which became the John Deere 30 series, were made between 1958 and 1960. The John Deere 435 with a 32 hp GM Detroit twin-cylinder two-stroke diesel engine was otherwise very similar to the 29 hp petrol-engined 430. The 430 and 435 were the last of the long line of twin-cylinder John Deere tractors.

John Deere was the only major North American tractor manufacturer without a factory in Europe until they acquired a major share in the Lanz factory at Mannheim in Germany in 1956. The Lanz Bulldog tractor, which had been made at Mannheim for thirty-five years, was discontinued and following a huge investment in the factory the first multi-cylinder John Deere-Lanz 300 and 500 tractors appeared in 1961. Meanwhile, in America the first of the new 10 series tractors which gradually replaced the earlier 30 series was launched in 1960. Wheeled and crawler versions of the John Deere 1010 and 2010 came first, the 3010 and 4010 were added in 1961 and the 125 hp 5010 appeared in 1963. Most British farmers were buying new diesel tractors in the early 1960s but John Deere petrol- and LPG-engined tractors were still being made for the North American market.

After being appointed the sole John Deere distributor for the UK in 1934 FA Standen at St Ives in Cambridgeshire sold a number of four-wheel and rowcrop tricycle-wheeled Model A tractors to Lincolnshire fen farmers. Changes in 1937 resulted in the appointment of H Leverton, Jack Olding, LO Tractors and FA Standen, now at Ely, as John Deere distributors in the UK. LO Tractors, jointly owned by Leverton and Jack Olding, were the John Deere agents in Scotland. The limited number of John Deere tractors supplied through the Lend-Lease programme dried up by 1945 and by 1951 all four UK distributors had relinquished their interest in John Deere equipment. Shortly after John Deere bought the Lanz factory at Mannheim H Leverton at Spalding was reinstated as their sole importer for the UK. This

164. Launched in 1964 the John Deere 4020 had a 100 hp six-cylinder diesel engine.

arrangement continued until Lundell GB took over the franchise in 1962 and within two years John Deere had bought the Lundell business at Edenbridge in Kent. Ford and Massey Ferguson had the lion's share of UK tractor sales in the early 1960s making it difficult for Lundell, who had been selling forage harvesters and some John Deere machinery, to break into the British tractor market. Lundell delivered some John Deere 4010 tractors in wooden packing cases to four of their dealers in 1964 and two of them – Ben Burgess at Norwich and P Tuckwell in Suffolk – are still selling John Deere tractors. The big green 4010 with its 93 hp six-cylinder engine was viewed with some apprehension by British farmers and apart from a John Deere mounted reversible plough there were very few implements big enough to make full use of this new-found power.

The John Deere 20 series with engines in the 47–143 hp bracket replaced the 10 series tractors over a period of three years during the mid-1960s. The 3020 and 4020 were launched in 1964, the 1020 and 2020 appeared in 1965 and the 1120 and 5020 completed the range in 1966. The 4020 accounted for almost half of John Deere's worldwide tractor sales when they opened a distribution centre at Langar in Nottinghamshire in 1966. Technically well ahead of its time, the 100 hp six-cylinder 4020 had an eight forward and two reverse synchro transmission, power steering, a two-speed power

take-off with separate 540 and 1,000 rpm shafts, a centre power shaft for a mid-mounted mower and an optional hydraulically operated diff-lock.

The 4020 was the first John Deere tractor with a ROPS (roll-over protective structure) safety frame and the first tractor on the British market with an optional eight forward and four reverse powershift transmission with a top speed of 20 mph. Like the Ford Select-O-Speed an inching pedal was provided for implement hitching and emergency stops. The powershift transmission was also optional for the 3020 with a four-cylinder 82 hp diesel engine. Petrol-engined 20 series tractors and the 3020 with an LPG engine were available in America.

Introduced in 1966 the 56 hp John Deere 710, made at Mannheim, was a direct competitor for the Ford 4000 and Massey Ferguson 165. The specification, included a ten forward and three reverse synchromesh gearbox, a 1,000 rpm live power shaft drive on one the side of the engine for a mid-mounted mower, two power take-off shafts at the rear, power steering and live top link sensing hydraulics.

The Mannheim-built 1020, 1120 and 2020 medium-powered tractors with at least 1,000 interchangeable parts did not appear in the UK until 1967. The 47 hp 1020 had a three-cylinder direct injection engine, the 53 hp 1120 and 64 hp 2020 had four cylinders and with many common features the basic prices were £899, £990.5s and £1,115.15s respectively. Closed centre hydraulics with lower link sensing, a hand lever- and foot pedal-operated diff-lock, hydraulic disc brakes, a front-mounted fuel tank and a foot throttle were standard. Optional extras included a dual speed power take-off, a Hi-Lo change-on-the-move speed change in any gear and power steering. The driving controls were

165. The 2140 was one of seven John Deere 'Schedule Masters' 40 series introduced in 1979.

grouped conveniently around the standard seat or an optional fully adjustable sprung version which added £9 to the price of the 1020. The 20 series was extended in 1969 with the launch of the 2520 and the 4520 and these tractors were made until the 30 series appeared in 1972. The 53 hp 1120 was retained until 1974 when the John Deere 1130 also rated at 53 hp replaced it.

The 2030 with a petrol or diesel engine launched in 1971 was the first of the new John Deere 30 series to be produced in Germany and America. The 30 series eventually ranged from the three-cylinder 48 hp 1030 for the small farm to the 215 hp articulated four-wheel drive 8430 and 275 hp 8630. Hydraulic trailer braking kits were offered for the 1130 and 3130 models and hydrostatic front-wheel drive was optional on the 4230 and the 4430. The 53–97 hp tractors had tilt-back safety cabs and the 122 hp six-cylinder 4230, also available in the UK, had a sixteen forward and six reverse gearbox, power steering and a safety rollbar. A re-styled and refined 30 series with an 'operator protection unit' or quiet cab, was claimed to surpass all safety and noise level regulations in force at the time. Sales literature explained that the Operator's Protection Unit (OPU) was 'a place to work in comfort, peace and quiet all day long without fatigue'. The leaflet continued that 'when sitting inside the John Deere OPU the elements can't touch you. The heater and windscreen de-froster with a fresh air intake, inside lighting, sunshield and rear-view

mirror are all standard and air conditioning will soon be available as an option.'

In 1977 John Deere announced four new American-built 40 series two-wheel drive tractors from the 100 hp 4040 to the 177 hp 4640, all with optional four-wheel drive. Seven 'Schedule Masters' 40 series tractors from the 50 hp 1040 to the 97 hp 3140 also made at the John Deere factory at Mannheim were added in 1979. The 'Schedule Masters' had an eight-speed synchromesh gearbox or optional Power-Synchron change-on-the-move high-low transmission and hydrostatic steering. Mechanical front-wheel drive controlled with a switch on the instrument panel was available for the 62–97 hp tractors. Prices ranged from £7,245 for the 50 hp 1040 to £15,178 for the four-wheel drive 97 hp 3140. The 4040 and 4240 with turbocharged 115 and 132 hp engines and a ZF mechanical four-wheel drive were added in 1981. A diff-lock, which automatically engaged when the wheels started to slip, was standard on the more powerful 40 series tractors.

New models appeared thick and fast in the early 1980s. Five 50–82 hp LP series tractors for livestock farmers and the XE 40 Economy range appeared in 1983. The 352 hp 8850 with a V8 turbocharged and intercooled engine was launched in the same year along with the 115 hp 4040S and 132 hp 4240S with sixteen speed Quad Range transmissions.

John Deere introduced five high horsepower 50 series tractors in 1982 and a year later a range of sixteen 50 series tractors with 44–352 hp engines had replaced the 40 series. Marketed as 'The Producers' the 50 series included four conventional four-wheel drive models starting with the 140 hp John Deere 4350 and two articulated four-wheel drive tractors including the flagship 8850 with a touch-control digital performance monitoring system in the cab. The smaller 50 series tractors were made at Mannheim and introduced to British farmers in 1986. The three-cylinder 44 hp 1550 was the only two-wheel drive model in the range, the 50 hp 1750 to the six-cylinder 100 hp 3350 were available in two- and four-wheel drive and the even more powerful tractors had four-wheel drive. A sixteen-speed Power-Synchron 40 kph transmission, standard on six-cylinder models, was optional for the other four-wheel drive 50 series tractors. Fuel savings of between 10 and 20 per cent and less engine wear compared with the earlier 40 series tractors were claimed advantages of the automatic eco-viscous drive radiator cooling fan used on the new tractors.

Six-cylinder turbocharged engines and a fifteen-speed powershift transmission were standard on the American-built four-wheel drive 128 hp 4050 and 144 hp 4250 which appeared in the UK in 1987. With category II and III linkage, a hydro-cushion seat and stereo radio in the cab and prices approaching £40,000 they were advertised as the first tractors in the UK to have a full power shift transmission with clutchless control in forward and reverse.

Five new 55 series four-wheel drive 'rowcrop tractors' with 128–228 hp six-cylinder engines and a fifteen forward and four reverse powershift transmission were launched in 1990. They had the John Deere 'Intellitrak Monitoring System' which not only monitored tractor performance and field operation but was also used by service engineers for diagnostic checks to locate faults. John Deere's rack and pinion rear axle adjustment for quick track width changes was re-introduced on the smaller 60 series introduced in 1991 and the more powerful 8560 and 8760 were improved with the addition of an electronic engine governor.

The unitary construction principle used for the best part of forty years was dropped when John Deere returned to a full-frame chassis for the engine and transmission on the 6000 and 7000 series launched in 1992. The four 6000 series tractors had 75–100 hp four-cylinder engines and the option of a SynchroPlus or PowerQuad shift-under-load transmissions. A nineteen forward and seven reverse speed Powershift transmission with a electronically controlled shifting was available at extra cost for the three 130, 150 and 170 hp six-cylinder 7000 series tractors. An improved hydraulic system had a pressure and flow compensating system (PFC) with a low standby pressure to help reduce fuel consumption.

166. The three-cylinder turbocharged John Deere 1950 – added to the 50 series in 1988 – was made until 1993.

167. An improved cab with storage for the farm computer was a new feature of the John Deere 6100 series launched at the 1997 Royal Smithfield Show.

John Deere tractors in 1993 included the 6000 and 7000 series, seven 50–88 hp 50 series models and the six-cylinder 4755 and 4955 rated at 190 and 228 hp respectively.

The driving controls, including a computerised headland management system on the 185–260 hp 8000 series, launched in 1995 were located on the Command arm rest attached to the swivel seat in the TechCentre cab. Other features included a sixteen-speed electro-hydraulically controlled powershift transmission with field cruise control and load sensing hydraulics. Three 7010 series with 140 hp, 155 hp and 175 hp engines replaced the 7000 series in 1996 when 84 hp and 100 hp models were added to the 6000 series. There was also a 55–85 hp 3000 series with an 18 forward and reverse speed transmission for the smaller farm. Standard versions of the 55 and 70 hp 5300 and 5400 in the new 5000 series launched in 1997 were assembled by Carraro in Italy for the European market.

The 6100 series introduced at the 1997 Royal Smithfield Show had new engines and an improved cab with storage space for the farm laptop computer. Four- and six-cylinder engines were used for the 75 hp 6010 to the 135 hp 6910 and the John Deere triple-link front

axle suspension system was optional on the 100 hp 6310.

Four John Deere 8000T series rubber-track crawlers with 185–260 hp engines based on the 8000 series wheeled tractors appeared in 1997. They had sixteen forward and five reverse transmission and friction drive tracks with hydraulic cylinders used to exert pressure on the front idler wheels and maintain track tension. The hydrostatic steering system had an electro-hydraulically engaged planetary final drive and the steering wheel varied the amount of power transmitted to each track when changing direction. Prices in 1998 for the 185 hp four-wheel drive John Deere 8100 started at £72,000 and the rubber-tracked 8100T cost £103,900. An improved five-model John Deere 6010 SE economy range with 75–105 hp engines and four transmission options, including a sixteen x sixteen PowerQuad option with four powershift gears, replaced the 6010 series in 1999. There was a choice of two- or four-wheel drive and standard or low-profile cabs.

John Deere had a busy year in 2002 when they launched the 5100, 6200, 8020 and 9020 series tractors. The three new 53, 74 and 80 hp John Deere 5100 series that replaced the earlier 5000 series were designed for the European market. Options for the 5100 series

168. The 8400T was one of the rubber-tracked versions of the wheeled John Deere 8400 series.

169. The six-cylinder 425 hp John Deere 9400 articulated tractor with a twenty-four forward and six reverse speed transmission was launched in 1999.

included two- or four-wheel drive, a twelve or twenty-four forward speed transmission with a top speed of 20 or 25 mph and a comfort or air suspension seat in a new roomy cab. Ten models in the new John Deere 6020 series, which superseded the 6010 series, ranged from the 80 hp 6120 to the 160 hp 6920S.

Optional equipment included the choice of various stepless transmissions, triple-link front suspension and a hydraulically self-levelling cab. Electronic engine governing on the 6420S provided up to 120 hp at the drawbar and an extra 5 hp at the power take-off.

The six-cylinder 8020 series ranging from the 200 hp 8120 to the 295 hp 8520 replaced the 8100 series in 2002. The new John Deere 9320, 9420 and 9520 articulated wheeled tractors, together with the 9320T and the 9420T rubber-tracked models with 375, 425 and 450 hp Power Tech engines, were added in the same year. An eighteen-speed automatic powershift transmission was used for the 9020 tractors and a twenty-four forward and six reverse gearbox was standard on the 9300T and 9400T rubber track models. The Mannheim-built John Deere 5020 series with 72, 80 and 88 hp engines with a sixteen forward and sixteen reverse PowerQuad transmission was added in 2003.

Three models of the 7020 series John Deere tractors introduced in 2003 had the latest Tier II compliant six-cylinder engines with common rail injection which provided an automatic transport power boost on steep inclines and when towing heavy loads. Various transmission options included the 40 or 50 kph twenty-speed PowerQuad Plus, AutoQuad Plus with manual or automatic gear changes and infinitely variable

AutoPowr. Other features included a new modular frame, larger restyled cabs with ActiveSeat suspension and optional front power take-off and hydraulic linkage. Five models of the new 8030 from the 225 hp 8130 to the 330 hp 8530 superseded the John Deere 8020 range in 2005. The flagship 8530 with AutoPowr infinitely variable and stepless transmission had electronic engine and transmission management and field cruise control.

Fifty years after John Deere acquired the Lanz tractor factory at Mannheim in order to establish a foothold in Europe there were at least fifty models of green and yellow tractor on the European market. The 2006 price list ranged from five 5015 series tractors with 55–88 hp engines and SynchroShift or power shuttle transmissions to the John Deere 9020 series of rubber-tracked and articulated four-wheel drive tractors. The basic specification for the 9020 series included 405–535 hp six-cylinder Powertech engines with four valves per cylinder, an eighteen-speed automatic powershift transmission and a cab full of computerised controls.

Chapter 5

Kendall – Loyd

Kendall

The tricycle-wheeled Kendall tractor was made by Grantham Productions in Lincolnshire in 1945 and 1946. Initially known as the Kendall-Beaumont the tractor with a 6 hp Beaumont three-cylinder radial engine was designed by Denis Kendall MP to reduce the cost of mechanised farming on small farms or to be used for light duties on larger farms. As the Kendall lacked power a supercharger was fitted but it was still underpowered and the radial engine was replaced with an 8 hp Douglas twin-cylinder air-cooled petrol engine. The Kendall was claimed to be equally suitable for light fieldwork at speeds of up to 6 mph and 20 mph for road haulage work. Grantham Tractors went into liquidation in 1947 and Newman Industries at Yate near Bristol bought the factory and business. The new owners refurbished the factory and in 1948 they introduced the very similar tricycle-wheeled Newman tractor with a Coventry Victor petrol engine.

Kubota

A large number of Japanese mini-tractors had been sold in America when the Marubeni Corporation tractor division established an outlet at Whitley Bridge in Yorkshire to sell Kubota tractors. The 24 hp three-cylinder L225, the two-cylinder 17 hp L175 and 12½ hp B6000 with water-cooled diesel engines were seen for the first time in the UK in 1975. The tractors had eight forward and two reverse gears, three-point linkage, front and rear power take-off and independent brakes. The 24 hp four-wheel drive L245 was added in 1977 and the three-cylinder 16 hp B7100 with six forward and two reverse gears followed in 1978.

Compact tractor horsepower gradually increased as the years passed. The four-wheel drive 34 hp Kubota L345DT launched in 1980 was the most powerful of a ten-model range. It had a dual clutch, eight forward and two reverse gears, power steering, front and rear power take-off and wet disc brakes. Having outgrown the Yorkshire premises Kubota moved to Thame in Oxfordshire

170. The first Kendall tractors had a three-cylinder radial engine.

171. Two- and four-wheel drive models of the Kubota B8200 were launched in 1983.

in 1982. New models were added over the years including the two- and four-wheel drive 19 hp B8200 and the B8200HST with hydrostatic transmission launched in 1983.

Kubota Tractors also made the two- and four-wheel drive M series tractors in Canada for the North American market in the 1980s. They included the M5950 and M7950 with 62 and 84 hp water-cooled engines, sixteen forward and eight reverse gears, hydraulic clutch, wet disc brakes and an independent power take-off with ground speed.

The four-wheel drive L5450 introduced in 1990 with a 54 hp five-cylinder engine, wet clutch, power steering and hydraulic shuttle transmission was the most powerful Kubota tractor yet seen in the UK. The two-cylinder 12½ hp four-wheel drive B4200 was the smallest model in the 1993 range which included the 29½ hp L2550 GST with a Glide Shift Transmission providing eight clutchless changes in speed and direction.

Three new L series Grandel compact tractors launched in 1994 had sixteen forward and reverse speeds in a synchromesh gearbox, a creep speed box and a hydraulic shuttle reverser. Kubota water-cooled indirect injection 34, 37 and 44 hp engines provided the power for the Grandel series with hydrostatic steering and wet disc brakes.

The 1997 Kubota catalogue included the 12½–24 hp B series with water-cooled engines, the 34–43 hp Grandel L models and the 29 hp four-wheel drive ST30 with hydrostatic transmission. The B series specification included a six forward and two reverse gearbox and mid- and rear-power take-off shafts. Optional equipment included bi-speed turning, hydrostatic transmission and either a safety frame or cab. The bi-speed turning system for four-wheel drive models increased the front wheel speed by about 60 per cent more than the rear wheels to give a much smaller turning circle.

172. High ground clearance was a feature of the four-wheel drive 44 hp Kubota Grandel L4200.

The Kubota ME series and GL30 series were both current in 2004. Three Kubota ME series tractors had 62–94 SAE hp engines, the smallest 58 hp 5700 DTQ had a five-cylinder engine and a twelve forward and reverse powershift transmission. The 87 and 94 hp M series had four-cylinder turbocharged engines and an eighteen-speed powershift transmission. The Kubota GL 30 series L3830, L4630 and L5030 tractors had 38–50 hp engines with the option of a Glide Shift (GST) transmission or fully synchronised main and shuttle (FST) transmission. The four-cylinder 105 hp ME105S with a thirty-two-speed powershift transmission was included in the 2006 ME series price list.

With an increasing interest in the farm tractor market Kubota attended the Royal Smithfield Show for the first time in 2004 when they exhibited their first 100 hp tractor. The four-wheel drive M105S with a four-cylinder turbocharged and intercooled Kubota engine had a sixteen forward and reverse synchronised gearbox doubled up by a Dual Speed shuttle, hydraulically engaged power take-off and dual level air-conditioning in the cab.

Lamborghini

Ferruccio Lamborghini made his first farm tractor in 1948 but within twenty years the Italian engineer was more famous for his high-performance sports cars. The 39 hp three-cylinder, diesel-engined 5C crawler introduced in the early 1960s was the first Lamborghini tractor to attract any serious interest outside Italy. It had a very unusual track design with three pneumatic tyred transport wheels that were used to raise the tracks off the ground when the tractor was taken on the public highway. The two rear transport wheels were driven by the track sprockets and were skid steered with the steering levers. A front castor wheel enabled the driver to make relatively sharp turns with the tractor. A narrow vineyard version of the Lamborghini 5C crawler had a standard clockwise power take-off shaft and a second anti-clockwise power shaft which ran at 2,000 rpm.

Lamborghini sold the tractor business to SAME in 1972 and by the mid-1970s Maulden Engineering at Flitwick in Bedfordshire had introduced a range of 38–105 hp Lamborghini tractors including the new 67

173. The Lamborghini 1R was made in the late 1960s.

hp Lamborghini 654 to British farmers. Competition for tractor sales was fierce in the late 1970s and there were some good deals around. A Vauxhall Cavalier car was offered as a free gift with the first three Lamborghini tractors bought and paid for in full at the 1980 Royal Show. As the car cost just under £4,600 it was equivalent to a 30 per cent discount.

There were eight Lamborghini two- and four-wheel drive tractors on the Maulden Engineering stand at the 1980 Royal Smithfield Show. The 105 and 125 hp six-cylinder 1056DT and 1256DT were the most powerful, followed by the 92 hp five-cylinder R955DT. The four-cylinder 62, 72 and 82 hp models with air-cooled engines had a twelve-speed synchromesh gearbox, the three-cylinder 53 hp R503 and 62 hp R603 had safety roll bars and three crawler models with 38, 59 and 60 hp engines completed the range.

The Lamborghini 1356 and turbocharged 1556, launched in 1981 with water-cooled engines rated at 135 and 155 hp, departed from the traditional air-cooled power units used for Lamborghini tractors. Sales literature explained that because of the advantages of improved fuel consumption and lower noise levels they had changed to water-cooled

174. The C553 crawler with a three-cylinder 59 hp direct injection engine and an eight forward and four reverse gearbox was one of a group of three Lamborghini crawlers on the market in the late 1970s. The C553 cost £6,610 and its optional cab added £995 to the price.

175. Maulden Engineering imported six models of Lamborghini tractors in the 69–140 hp bracket. They included the 100 hp R 955 with 100 hp under the bonnet in 1980.

engines for their high horsepower tractors. However, four- and six-cylinder air-cooled engines were used for the new 75 hp C754L crawler based on the 854 wheeled model and the 115 hp Lamborghini 1156 wheeled tractor also launched in 1981.

Universal Tractors at Brough in Yorkshire, who later became Linx Agriculture, were appointed concessionaires for Lamborghini tractors in 1983. The 60–155 hp two- and four-wheel drive tractors had a pedal-operated hydraulic trailer braking system and oil-immersed front disc brakes were standard on four-wheel drive models over 110 hp. When Lynx Agriculture went into receivership in 1986 UK distribution of the Italian tractors was taken over by SAME-Lamborghini at Barby in Warwickshire. There were nine two- and four-wheel drive models with water-cooled engines, ranging from the new 60 hp Lamborghini 660 to the well-established 165 hp 1706. Lamborghini and SAME tractors were very similar apart from their colour and the air- or oil-cooled engines used for the SAME range.

176. Lamborghini tractors imported by Lynx Agriculture in the early 1980s included the 135 hp 1356DT with hydrostatic steering and a cab that was supported on four silent blocks.

177. Introduced in 1992 the 165 hp Lamborghini Racing 165 had a 27-speed powershift transmission.

178. An optional reverse drive with twenty-four forward and reverse speeds was available for the early 1990s three-cylinder 75 hp Lamborghini 775-F Plus.

'Drive by wire' was the 1989 advertising slogan for the new four-wheel drive Lamborghini Grand Prix 674-70, 774-80 and turbocharged 874-90 with more than a hint of motor racing in their name. They were advertised as the first tractors to have an individual electronically governed injection pump for each of the four cylinders. The 70, 80 and 90 hp Grand Prix tractors had push-button control for the four-speed power take-off and the 90 hp model had a forty forward speed gearbox. Lamborghini sports car stylists had some influence in the design of the bonnet and cab on the Formula 115 and 135 tractors launched in 1990. The six-cylinder 115 and 132 hp tractors had an electronic engine monitoring system and a thirty-six forward and reverse speed gearbox. The reverse drive Twin Systems 70 version of the 70 hp Lamborghini 700 with modified driving controls, front hydraulic linkage and front power take-off was launched along with the reverse drive SAME Dual Trac 70 at the 1990 Royal Show.

The Lamborghini Racing 165 introduced in 1992 with electronic engine governing and hydraulic

179. An electronic engine management system was a feature of the six-cylinder 105 hp Lamborghini Premium 1060.

linkage control had a clutchless twenty-seven-speed powershift gearbox and with the exception of its water-cooled engine the Racing 165 was very similar to the SAME Titan 160.

When the Racing 190 with a turbocharged six-cylinder engine was launched later in 1992 Lamborghini tractors ranged from the 25 hp Runner 250 to the 189 hp Racing 190. The 150 hp Lamborghini Racing 150 launched in 1994 had a SAME clutchless transmission with nine powershift speeds in each of three ranges selected electronically with a push-button switch on the joystick control lever.

An agreement with AGCO in the early 1990s resulted in

180. The 80 hp Lamborghini 774-80 Grand Prix cost £21,750 in 2004.

some Lamborghini tractors being sold in America under the White brand name with silver and black paintwork. Power take-off horsepower was an important factor for American farmers when buying a tractor. The 60–105 pto hp Lamborghini models with the White colour scheme were added to the more powerful range of White tractors in the 120–200 pto hp bracket made by AGCO at Independence, Missouri.

The 85–105 hp Lamborghini Premium models were launched at the 1994 Royal Smithfield Show. An electronic engine governor, standard on the six-cylinder Premium 1060, was optional on the other models in the range which had SAME-Lamborghini electronic hydraulic linkage with lower link sensing and there was a choice of a mechanical or electronic twenty forward and reverse powershift transmission.

SAME Deutz-Fahr included twenty-three Lamborghini models from the 25hp Runner 250 to the flagship Racing 190 with a six-cylinder engine and a twenty-seven forward and reverse speed transmission in the 1998 UK price list. The Champion 120, 130 and 150 launched at that year's Royal Smithfield Show were more sophisticated versions of the medium-powered SAME Rubin tractors.

The Sprint and Grand Prix, the 60–83 hp Agile models, the 230 and 260 hp Victory and the new 91–205 hp R series were included in the 2004 Lamborghini catalogue. The Agile and Sprint tractors with three- and four-cylinder engines were available with a thirty-five forward and fifteen reverse or an optional three range forty-five forward and reverse gearbox. Options for the four- and six-cylinder 100 hp plus R series included a fifteen forward and reverse synchro-shuttle and a forty, fifty-four or sixty forward and reverse speed power shuttle transmission.

Landini

Giovanni Landini was making steam engines and vineyard machinery at his factory at Fabbrico in Italy in 1884. His first tractor, with a single-cylinder two-stroke 'hot bulb' semi-diesel engine, appeared in 1910. Several tractor manufacturers including Deutz, Landini, Lanz and Munktell also used two-stroke hot bulb engines for the next forty years or so. Landini's three sons continued the business when he died in 1924 and in 1925 they introduced the Landini 25/30 tractor with a 30 hp horizontal two-stroke semi-diesel engine. A 40 hp Landini tractor appeared in 1930 and the 48 hp single-cylinder water-cooled Super Landini with three forward gears and one reverse introduced in 1934 was made until the outbreak of war in 1939. The 30 hp Landini Velite appeared in 1935 and when the Buffalo was added in the late 1930s three Landini models with semi-diesel engines were being made in Italy.

Semi-diesel hot bulb engines, which ran on low-quality fuel, were still used when Landini tractor production was restarted with the launch of the 25–30 hp L25 and 30–40 hp L35 in 1950. The 45–50 hp Landini L45 with six forward gears and one reverse was added in 1952 and the 55–60 hp L55 appeared in 1954. Optional Roadless

181. The single-cylinder hot bulb Landini L45 on trial in 1952 with Roadless DG half tracks.
(Stuart Gibbard)

half tracks were available for the L25, L35, L45 and L55 tractors. The Landinetta, launched in 1957 with the engine and transmission offset in the style of the Allis-Chalmers Model B, was the last model of Landini semi-diesel engined tractor made at Fabbrico.

Following an agreement with Perkins in 1957 Landini made the Peterborough-designed diesel engines under licence in Italy. They included a three-cylinder 30 hp diesel engine for the Landini R35 launched in 1957. The same engine was used for the C35 crawler introduced in 1959 and a four-cylinder 50 hp engine built to a Perkins design provided the power for the six forward and two reverse speed R50 with optional four-wheel drive. The

182. The Perkins-engined Landini Trekker range of crawler tractors was launched in 1996.

Landinetta was superseded by the R25 with a 28 hp Perkins engine in 1959, the tractor was re-badged as the R3000 in 1960 and Massey Ferguson, who also owned Perkins, bought the Fabbrico factory the same year.

Landini tractor design moved forward under Massey Ferguson and from the early 1960s blue Landini and red Massey Ferguson tractors came off the same production line. The four-cylinder Landini R4500 based on the Massey Ferguson 65 and introduced in 1962 was the first model made at Fabbrico under the new ownership. The 45 hp Landini C4500 crawler was added in 1963. The Massey Ferguson influence was still obvious when the 80 hp two-wheel drive R8000 and the four-wheel drive DT8000 were launched in 1967. The DT (Dual Traction) model with a Selene front axle was designed on the drawing board as a four-wheel drive tractor rather than as a two-wheel drive with a live front axle added at a later stage. Similar Selene front-wheel drive conversion kits for Massey Ferguson 165, 175 and 185 tractors were marketed by Four Wheel Traction Ltd.

A new range of Landini crawlers introduced in 1975 included the 41 hp three-cylinder C6500 with eight forward and four reverse gears with the option of

standard and narrow width tracks. The Landini C6500 also appeared at the 1976 Royal Show in Massey Ferguson colours as the MF 134C. The 47 hp MF 154C and 61 hp 174C crawlers were also made by Landini and the four-wheel drive 14500DT with a 107 hp Perkins A 6.354 engine launched the same year was the first Landini tractor made at Fabbrico with more than 100 hp under the bonnet.

Landini were still a wholly owned subsidiary of Massey Ferguson when they made the first specialist Landini fruit tractors in 1982. Vineyard models were added in 1986. The prefix letter V denoted a narrow vineyard model, F was used for standard orchard tractors and L for wide orchard models. Almost half of the 12,000 or so tractors made at Fabbrico in the late 1980s were painted in Massey Ferguson colours until the Varity Corporation sold Massey Ferguson to AGCO in 1990 which meant that at last Landini were able to sell their blue tractors in Britain. A Luxembourg company owned the majority share of Landini but Massey Ferguson retained an interest in the business.

The Landini tractor range in 1990 included the 80 and 10,000 series wheeled tractors and 47–80 hp crawlers. The two- and four-wheel drive 80 series included the Landini 6880, 7880, and 8880 and 9880 with 62, 71, 80 and 93 hp four-cylinder Perkins engines and a twenty-four forward and twelve reverse synchromesh gearbox with an optional ten forward speed creeper box. Hydraulic inboard multi-disc brakes and diff-lock were standard. Optional Landtronic electronic draft control and performance monitor were added later in the year. The 80 series transmission system and lower link sensing hydraulics were also used for the Landini 10,000 series with 103, 110 and 132 hp Perkins engines.

British farmers were faced with a choice of more than 500 different models of tractor from thirty manufacturers and importers in 1991 when Watveare Ltd, who were already marketing Deutz-Fahr and MB-Trac tractors became the UK distributor for Landini. The Italian tractor range at the time included the four-wheel drive six-cylinder 103 and 132 hp Landini DT1000 and DT14500 and both two and four-wheel drive versions of the 6680, 7880, 8880 and 9880. High-clearance models of the four-wheel drive Landini 8880 and 9880 with equal-sized wheels and disc brakes on all four wheels were also available in the UK and the Viewmaster 7880 and 8880 tractors with sloping bonnets were added in 1992. The two- and four-wheel drive 62, 71 and 80 hp Landini Blizzard tractors with a twelve forward and reverse speed gearbox suitable for stock and mixed farms appeared in the same year. Forward control versions of the 8880 and 9880 with a rear platform for a sprayer or fertiliser spreader were added in 1993.

The ARGO group acquired the majority Landini shareholding from AGCO in 1994 when Landini AMS at Bury St Edmunds in Suffolk became the UK distributor for the light blue Italian tractors. The 110 and 123 hp Landini Legend DT115 and DT130 tractors, introduced to Italian farmers in 1994, were not available in the UK until 1996. The Massey Ferguson factory at Beauvais in France supplied some components for the six-cylinder Perkins-engined Legend tractors which had a thirty-six speed change-on-the-move transmission with a splitter and reverse shuttle. An optional creep speed box doubled the ratios to give the Legend seventy-two gears in each direction.

ARGO introduced the turbocharged 90 hp Blizzard DT95 with a twenty-four forward and twelve reverse 40 kph transmission and disc brakes on all four wheels in 1995 when they bought Valpadana which made compact tractors in Italy. A 60-degree steering angle combined with a fast-run front axle which increased front wheel speed when turning on the headland was a new feature on the four-wheel drive Globus 50, 60 and 65 tractors introduced in northern Europe in 1996. The Globus had a four-speed power take-off with reverse drive and the choice of a twelve, fifteen or twenty-five speed reverse shuttle transmission. Landini, who were selling Massey Ferguson tractors to Italian farmers in the late 1960s, made crawlers and fruit tractors for Massey Ferguson and Globus tractors which were sold in Iseki livery in Japan.

Landini were still building specialist tractors for Massey Ferguson in 1996 when they launched three models of the 'Trekker' crawler with 71, 85 and 95 hp Perkins engines. Features included a sixteen forward and eight reverse overdrive transmission, a hydraulic track tensioning system and category II hydraulic linkage. The Globus 50, 60 and 70 with Perkins 47, 56 and 63 hp engines respectively and the choice of a twelve, fifteen or twenty-five forward and reverse creep speed transmission and a three-speed power take-off appeared in the same year.

Motokov UK were appointed Landini distributors for the British Isles in 1998 and imported the Italian tractors in the 47–100 hp bracket. The 56 and 66 hp Landini Globus models appeared at the 1998 Royal Smithfield Show and the Landini Discovery tractors were added in 1999. There was an option of a fifteen or twenty-five speed shuttle gearbox for the four-wheel drive Globus tractors and the Deutz-engined 65 and 85 hp Discovery models with equal-sized wheels had a thirty-six forward and reverse shuttle transmission and oil-cooled disc brakes. The optional reverse drive for the Discovery series made it possible to convert the tractor from forward- to rear-facing operation in a matter of seconds.

Further improvements were made to the 50–100 hp Landini range in 2000 and a new generation of Landini Legend tractors appeared in 2001. The new 80–100 hp Vision line appeared in 2002 when marketing came under the control of Landini UK based at the McCormick factory at Doncaster. Landini Mistral, Globus, Legend, Mythos, Powerfarm, Vision and

183. Five models of the Landini Legend range with 118 to 183 hp Perkins engines were current in 2003.

Starland tractors with engines in the 50–270 hp bracket were current in 2004. Transmissions with up to 108 forward speeds and thirty-six in reverse with a top speed of 50 kph were available for the Legend series. The 85–105 hp Perkins-engined Vision range was available with a forty speed forward and reverse Power Shuttle or a synchro gearbox with twenty forward and reverse ratios.

The 270 hp Landini Starland D270 with a six-cylinder engine and an eighteen forward and nine reverse power shuttle gearbox made by Buhler Versatile in Canada, which cost £87,500, was the most expensive tractor in the 2004 Landini price list. A range of five utility Powerfarm tractors launched at that year's Royal Smithfield Show was designed for livestock farms and fruit or vegetable holdings. The Powerfarm models had a three- or four-cylinder 59–99 hp Perkins engine, a twelve forward and reverse or a twenty-four forward and twelve synchro shuttle gearbox and two-speed power take-off. Optional extras included two- or four-wheel drive, a twelve forward and twelve reverse creep speed gearbox and a safety cab or a 'quiet open platform' for working in polytunnels.

Lanz

Heinrich Lanz founded an engineering company at Mannheim in Germany in 1859 and within a few years he was selling British-built Clayton & Shuttleworth threshing machines to German farmers. Chaff cutters and barn machinery were made at Mannheim by 1867 and the early 1880s Lanz products included steam engines and threshing machines. The Lanz Landbaumotor with four-stroke 80 hp engine and a mounted chain-driven rotary cultivator which could be removed to leave the tractor free for other work was probably the first agricultural machine with a crude form of hydraulic lift.

The first Bulldog or HL Landbaumotor (Heinrich Lanz agricultural engine) made in 1921 had a 12 hp single-cylinder, horizontal semi-diesel engine. The two-stroke hot bulb engine developed by Dr Fritz Huber, an engineer at the Lanz factory in Mannheim, would run on almost any type of liquid fuel including paraffin, creosote, palm oil and ethanol. Lanz tractors were known as Bulldogs throughout their forty-year production run because of the vague resemblance of the hot bulb on the cylinder head at the front of the tractor to the face of the British bulldog. Variants included the Field Bulldog, Highway Bulldog and the Express Bulldog with a cab for road haulage. The Bulldog was started by heating the hot bulb with a blowlamp before cranking the engine with the steering wheel inserted in the centre of the flywheel. The engine ran at 420 rpm and a direct chain drive from the flywheel to the rear wheels was engaged with a lever-operated clutch. The tractor had a top speed of 3½ mph and was reversed by stopping the engine and restarting it again with the flywheel running in the opposite direction.

Several Bulldog models appeared in the 1920s and 1930s. They included the HL, which with the engine speed increased to 500 rpm, was rated at 15 hp. The four-wheel drive Bulldog HP based on the HL with the front wheels larger than the rear wheels was added in 1923. The HP was an early example of a steel-wheeled articulated tractor but it was expensive and was only made for three years. The more powerful 22–30 hp HR2 introduced in 1929 had four forward and four reverse gears and like the HL the engine had to be stopped and restarted backwards to reverse the tractor.

The 15–30 hp Lanz Bulldog HR5 made between 1929 and 1935 had a pedal-operated clutch and a three forward speed and one reverse gearbox. There was also a rowcrop model and tractors supplied with optional pneumatic tyres had a higher top speed. Like earlier Bulldogs the HR5 engine had a dry crankcase with oil pump lubrication and the instruction book reminded drivers that the bearings needed to be lubricated by hand before starting the engine if the tractor had not been used for two days or more. The 38 hp Bulldog HR5 had a six-speed gearbox and sprung front axle. The engine could be set to run at 540 or 630 rpm, with the higher speed providing an additional 8 hp.

The D8500 and D9500 also introduced in 1934 had

184. The first Lanz Bulldog tractors were made at Mannheim in Germany in 1921.

the same engine which could be set to run at 540 and 630 rpm and developed 35 and 54 hp respectively. Other D series models, including the D 1560 crawler version of the Bulldog, appeared from time to time. Although it was steered in the usual way with two levers a steering wheel on a short shaft, stored at the back of the tractor, was only used to start the engine!

The semi-diesel engined Bulldog 06 series were made at Mannheim from the mid-1930s and more than 100,000 Bulldog tractors had been built when the factory was virtually destroyed in the early part of the World War Two. Some Bulldog D7006 and D9006 tractors made during the early part of the war ran on wood gas from a generator mounted on the tractor. Lanz struggled to survive in the immediate postwar years and although some of the more powerful 06 series tractors were made in the late 1940s the smaller Bulldogs did not appear until 1950. The first two figures of the 06 series model numbers indicated maximum horsepower and alternative engine speeds of 850, 950 and 1,050 rpm enabled the driver to vary the horsepower according to need. The D1706, D1906 and D2206 had a 130 mm bore and a 170 mm stroke engine while a 150 mm bore and 210 mm stroke engine was used for the D2806, D3206 and D3606. A six forward and two reverse speed gearbox and pneumatic tyres were standard features of Lanz 06 series tractors.

The D5506 introduced in 1950 was the first new Bulldog tractor to be made at Mannheim after the war. The hot bulb was moved to the left-hand side of the engine, a blowlamp was still an important part of the toolkit but an optional electric starter made life easier for the driver. The early 1950s 45 hp Lanz Bulldog D9506 had a single-cylinder two-stroke hot bulb engine and a six forward and two reverse gearbox with a top speed of 11 mph. Optional equipment included a 12 volt electric starter and an all-weather cab with side and rear curtains and celluloid windows. The hot bulb disappeared in 1952 when a flat-topped cylinder head with an ignition chamber was used for the semi-diesel engine for the D5506. The single-cylinder Bulldog breed finally disappeared in 1955 when Lanz launched the D1616 and D2016 tractors with full diesel engines.

The Lanz Alldog A1205 tool carrier, introduced in 1951, had a 12 hp air-cooled single-cylinder petrol engine, a five forward and one reverse gearbox and final drive housing at the rear of an open tool frame. The A1205 had been superseded by the Alldog A1305 with a single-cylinder 13 hp diesel engine when H Leverton & Co at Spalding demonstrated it in Britain for the first time at the 1952 sugar beet harvesting demonstration. Features of the Alldog included a power-take off shaft and hydraulic linkage at both ends, rear-mounted belt pulley and independent brakes. An advertisement for the Alldog, with a power take-off and hydraulic linkage at the front and rear, independent brakes and a rear-mounted belt pulley, claimed that the driver 'could attach a wide range

185. This 1935 vintage Lanz Bulldog with a semi-diesel hot bulb engine is typical of the 200,000 plus tractors made by Heinrich Lanz at Mannheim.

186. The hot bulb on the Lanz cylinder head, similar to the facial features of a bulldog, is said to have given the tractor its Bulldog name. The bracket attached to the front axle supported the blowlamp used to pre-heat the hot bulb before starting the engine.

165

187. The steering wheel was used to turn the flywheel when starting a Lanz Bulldog.

188. An optional 12-volt starter was available for the 45 hp Lanz Bulldog D9506 made at Mannheim in the early 1950s.

of front-, mid- and rear-mounted implements without difficulty'. The Alldog was front-wheel steered but the rear axle, connected to the tool frame by a single pivot pin, could be steered independently to reduce side-slip when working across slopes. The A1806 Alldog with an 18 hp water-cooled MWM diesel engine and a five-speed gearbox introduced in 1956 remained in production until the first John Deere-Lanz tractors were made at Mannheim in 1960. Equipment for the Alldog included front-, underslung and rear-mounted toolbars with a full range of tines and hoe blades, a mid-mounted plough and a single row sugar beet harvester.

John Deere had been looking for manufacturing facilities in Europe and they bought a majority share holding of Heinrich Lanz AG in 1956 and completed the acquisition of the

189. A flat-topped cylinder head replaced the hot bulb in 1952 when Lanz introduced the Bulldog 5506.

190. The Lanz Alldog tool carrier was used for rowcrop work and general haulage around the farm.

191. The John Deere-Lanz 500 with a four-cylinder water-cooled 36 hp diesel engine and a ten forward and three reverse gearbox launched in 1961 was the first of the new generation of Mannheim-built John Deere tractors.

Mannheim factory in 1960. The 11 hp single-cylinder two-stroke diesel-engined Lanz D1106 'Bulli' with a six forward and two reverse gearbox was made between 1956 and 1958. This blue tractor with red wheels was very different to the traditional Bulldog but it did not enjoy the same success. Bulldog-type tractors, including the 42 hp single-cylinder 4016 with green and yellow paintwork, were made on a limited scale at Mannheim.

The three-cylinder John Deere-Lanz crawler was introduced in 1959 and four-cylinder 28 and 36 hp John Deere-Lanz 300 and 500 tractors made at Mannheim were unveiled in 1961. The new green tractors with yellow wheels had independent coil spring suspension on the front axle, disc brakes and three-point linkage. The Lanz 300 and 500 had separate 540 and 1,000 power take-off shafts placed side by side at the rear and a third power shaft on one side of the tractor for a mid-mounted mower. The Lanz name disappeared in 1965 when the first John Deere 1020, 1120 and 2020 tractors were made at Mannheim.

Leyland

British Motor Holdings, which made Nuffield Tractors at Bathgate in Scotland, and the Leyland Motor Corporation merged in 1968 to form the British Leyland Motors Tractor Group. Production continued

at Bathgate and three 'new' two-tone blue Leyland tractors were introduced at the 1969 Royal Smithfield Show. The Leyland 154, 344 and 384, advertised as 'three new tractors for the seventies', had a new colour scheme but were otherwise very similar to the earlier Nuffield 4/25, 3/45 and 4/65.

The 154 was retained when the Leyland 245, 255, 270, 285/485 and 2100/4100 were introduced in 1972. The first digit in the model number denoted two- or four-wheel drive and the remaining numbers indicated horsepower. The six-cylinder 85 and 100 hp tractors were well equipped with ten forward and two reverse gears and a top speed of 21 mph. The specification included live hydraulics, dual speed independent power take-off, hydrostatic steering and multi-plate wet disc brakes. The medium-powered Leylands had four-cylinder engines, ten forward and two reverse gears, independent power take-off and dry disc brakes. Power steering was an optional extra.

The 262 and 272 replaced the 255 and 270 in 1976 and with the exception of the 154 all Leyland tractors had hydrostatic power steering and a 'Q' cab. A four-wheel drive conversion based on the Leyland 270 with a drive shaft running forward from a transfer box behind the main gearbox was sold as the Four Wheel Traction 270 in the mid-1970s. Prices were in an upward spiral at

192. The Leyland 154 with its new blue colour scheme was very similar to the earlier Nuffield 4/25.

the time when the Leyland 154 cost £2,392 compared with the £585 price tag for the similar BMC Mini-tractor in 1965. Leyland also broke the £10,000 barrier in 1976 when they advertised the four-wheel drive 4100 at £10,178. The Leyland 272H, added to the range in 1977, was a variant of the standard tractor with a high ratio final drive providing a 27 per cent speed increase in all forward gears.

The 245, 262/462 and 272/472 Leyland Synchro range launched in 1978 had a nine forward and three reverse synchromesh gearbox. The 272 and 472 engines were turbocharged from 1979 and with an extra 10 hp they became the 282 and 482. The Leyland 285 was improved with a nine forward and three reverse synchromesh gearbox in the same year.

The 154 had been made for ten years when the Leyland 302 superseded it in 1979. The new model, manufactured in Turkey, was similar to the 154 with a new more powerful engine and an improved hydraulic system.

A new range of two- and four-wheel drive Leyland tractors with harvest gold and black paintwork was launched at the 1980 Royal Smithfield Show. Unlike the earlier Leyland tractors the model numbers' first two

figures were now only a rough guide to horsepower and the 2 or 4 denoted two- or four-wheel drive. The four-cylinder 302 and the three-cylinder 502 with direct injection engines had a nine forward and three reverse synchromesh gearbox and disc brakes. The 602/604, 702/704 and turbocharged 802/804 had the same gearbox with a dual clutch, live hydraulics and two-speed power take-off. Depending on the price paid the driver could sit in the standard QM cab or optional Explorer cab with windscreen wiper and indicator switches on the steering column. An advertisement explained that the new Explorer cab provided 'a luxurious working environment which is second to none'.

The 92 hp 904 XL, added to the range in 1984, had a fifteen forward and five reverse gearbox which was optional for the 62, 72 and 82 hp tractors. Leyland reported a 45 per cent increase in sales for the new tractors compared with the previous range and the 302, which superseded the 235, was Leyland's answer to the growing numbers of Japanese compact tractors on the UK market. The quiet cab on the 302 could, after removing four bolts, be lifted off to work in low buildings or orchards and the bonnet tipped forward to service the engine.

193. Hydrostatic steering was standard on the Leyland 272 which superseded the 270 in 1976.

194. The 462, one of the new Leyland Synchro range tractors launched in 1978, had nine forward and three reverse gears.

195. Two- and four-wheel drive models of the 87 hp Lely 90 Hydro were made in the early 1970s.

The state-owned British Leyland was the first company to be privatised by Mrs Thatcher's government and Charles Nickerson, owner of Track-Marshall Ltd, bought the Leyland tractor division in 1982. Production of the five Leyland models from the 17 hp 302 to the 82 hp 802 now with a Marshall badge on the bonnet was transferred to Gainsborough and the completed tractors were given a thorough pre-delivery inspection.

Lely

Founded in 1948, implement makers Lely Industries at Maasland in Holland entered the tractor market in 1970 with the Hydro 90. The six-cylinder MWM direct injection diesel-engined tractor had a hydrostatic transmission with a stepless variable forward speed of up to 12½ mph and 7½ mph in reverse. Planetary final drive reduction gears transmitted the power to the wheels. The tractor had a diff-lock, rear drum brakes and the parking brake was automatically applied when the transmission was in neutral. Other features included hydrostatic steering, ZF hydraulic linkage and an 80 hp hydraulically driven dual-speed power take-off.

The Lely Hydro 150 with a 152 hp Ford engine, a high/low ratio hydrostatic transmission and steering, hydraulic brakes and category II hydraulic linkage was also made in the early 1970s. It was an early version of a 'systems tractor' with reversible driving controls and seat which could be 'driven with equal ease in either direction'.

Lely also made a prototype forward control tractor in 1972. The 178 hp Supertrac had a Perkins diesel engine, ten forward gears, a rear load-carrying platform and hydraulic linkage. A set of matched implements was planned but the project was not progressed and Lely concentrated on the development of their other products.

Lely returned to the tractor scene in 1979 when they set up a partnership with Iseki to market 13–35 hp Japanese TS and TX compact tractors in the UK. The Lely Iseki partnership continued until 1986 when Iseki established their own distribution company in Cambridgeshire.

196. *The Loyd Dragon was expected to cost £1,450 ex-works when it was announced in 1950.*

Loyd

Vivian Loyd at Camberley in Surrey made tracked military vehicles during the war years and they used this expertise in the late 1940s to manufacture agricultural tracklayers. The 33 hp Loyd DP with a water-cooled Turner V4 diesel engine and the 50 hp Loyd D with a Ford V8 petrol engine were the first Loyd crawlers. Both tractors had a four forward speed and one reverse gearbox, differential brake steering and military-type carrier tracks. Another version of the Loyd crawler with a Dorman diesel engine was made for a short while before the Dragon with conventional four bottom roller tracks, supplied by John Fowler Ltd, was launched at the 1950 Royal Smithfield Show.

Sales literature explained that the Loyd Dragon had been put through a series of tests including trials on an army tank track where potential distributors from the UK and overseas said they had 'never before seen a tractor subjected to such a destructive test'.

A Turner V4 or a Dorman-Ricardo in-line diesel engine both rated at 36½ hp was used for the Dragon which had a four forward and one reverse gearbox, multi-plate clutch and brake steering and power take-off. It had a maximum pull of 7,000 lb at the drawbar, a ground pressure of 6 lb per square inch was claimed and it weighed just over 3½ tons. In the event very few Loyd Dragon crawlers were built and the last one was made in 1952.

Chapter 6
Mailam – OTA

Mailam

An Italian Ford tractor dealer that also made industrial crawlers and bulldozers used Ford engines and gearboxes and various bought-in components including the tracks to build Mailam agricultural crawler tractors. The Mailam 5001, introduced to Italian farmers in 1966, had a 65 hp Ford 5000 engine, an eight forward and two speed reverse gearbox and multiple disc clutch and brake steering. A 75 hp Ford Force 5000 engine was used from 1968 for the Mailam 5001 tractor. A 100 hp six-cylinder Ford Industrial engine and a stronger transmission were used to build a bigger Mailam crawler with a power take-off and an optional belt pulley attachment for the power shaft.

Hertfordshire Ford tractor dealer Gates at Baldock imported three Mailam 5001 crawlers in 1970 and added a power take-off and hydraulic linkage to the tractors. The tractors were demonstrated at various locations mainly in Bedfordshire and Hertfordshire but farmers were not very impressed by the Italian crawler. Gates also imported some Mailam crawlers with an International Harvester TD9 transmission and Burford running gear. Little more was heard of Mailam crawler tractors and the Italian company ceased trading in the early 1970s.

Marshall

The Britannia Works at Gainsborough in Lincolnshire founded by William Marshall in 1848 was for many years the home of Marshall portable steam engines, traction engines and threshing machines. The first Marshall 'oil-engined' tractor was made at Gainsborough in 1906 and like other tractors of the day it had chain drive to the rear wheels. Marshalls also made a 70 hp Model F oil-engined tractor which was almost as big as a traction engine.

The 15/30 introduced in 1930, which weighed over three tons, was the first Marshall diesel tractor. Similar to the Lanz Bulldog tractor it had a single-cylinder two-stroke, 550 rpm horizontal engine with an 8 in bore and a 10½ in stroke. The 18/30, an improved version of the 15/30, was made between 1932 and 1934. Initially customers were able to specify the colour of their new Marshall tractor but from the mid-1930s the usual colour was green. Marshall's acquired Clayton & Shuttleworth in 1930 and this led to an early example of

197. The Mailam 5001 crawler was based on the Ford 5000 tractor. (Stuart Gibbard)

'badge engineering' when they sold a number of Marshall tractors in Belgium and Greece as the Clayton 18/30. Norfolk farmer Ben Burgess formed the Crude Oil Tractor Co in 1931 (later Ben Burgess Tractors) in Norwich, to sell the Marshall 18/30 with a £333.6s.8d price tag. He recalled that one customer had told him that as his horses walked at 3 mph he considered the 18/30's top speed of 3½ mph too fast for satisfactory ploughing!

198. The single-cylinder Marshall 12/20 diesel tractor was introduced in 1936.

British farming was sliding into a deep depression when the Marshall 12/20, still with the Lanz Bulldog influence, was sold to a Norfolk farmer in 1936. Two years later an improved 12/20 became the Marshall Model M.

The 1930s agricultural depression had a serious effect on tractor sales and as a result iron manufacturers Thomas Ward Ltd at Sheffield absorbed Marshalls in 1936 and changed their name to Marshall Sons & Co Ltd. Tractor engine speeds were gradually increased in the 1930s, the Marshall 15/30 ran at 550 rpm, the 12/20 had a top speed of 680 rpm and the Model M engine ran at 700 rpm.

A few Model M tractors were made during World War Two when the Marshall contribution to the war effort included midget submarines. Tractor production resumed at the Gainsborough works when the new Series I Field Marshall was launched in 1945. The new tractor had the same horizontal single-cylinder two-stroke diesel engine as the Model M but with its speed increased to 750 rpm the tractor developed 40 hp. It had three forward gears and one reverse, a diff-lock, transmission brake and there was a choice of steel wheels or pneumatic tyres. The mid-Brunswick green Field Marshall with silver wheels had a large external flywheel at one end of the crankshaft and a belt pulley combined with a cone clutch and its housing at the other end. An offset power take-off shaft and a winch were available at extra cost. There were two versions of the Series I Field

Marshall. The Mk I with a top speed of 6 mph was for farm work and the Mk II with more efficient brakes and a top speed of 9 mph was mainly used by threshing contractors.

The Field Marshall was started by hand with the aid of a decompression mechanism operated by a disc running in a spiral groove on the flywheel and when the disc ran out of the groove the engine returned to full compression. A smouldering paper wick in a metal rod, which was screwed into the cylinder head, provided some heat to help start the engine. On very cold days, the engine could be started with a special cartridge placed in a holder above the piston and fired with a suitable blunt instrument.

An advertisement claimed the Field Marshall would 'run for one-sixth of the cost of a petrol- or paraffin-engined tractor and plough an acre in an hour on one gallon of diesel fuel'. It was also pointed out that the tractor had only sixty-six working parts compared with an average of 194 on petrol tractors and would therefore only 'go wrong a third as often'! Low-cost maintenance was claimed as another plus but 'in return for its labour the tractor requires de-carbonising once or twice each year and the simple design of the engine enables unskilled labour to complete this task in three hours'. A sales leaflet explained that, 'Tests have shown cylinder wear to be only fifteen thousandths of an inch after three years' service and when necessary it is cheap and easy to have a new cylinder block fitted and start all over again'.

John Fowler of Leeds, like Marshall Sons & Co, became associated with Thomas Ward Ltd in 1946 and an improved Series 2 Field Marshall replaced the Series 1 in 1947. The new model did not have a diff-lock but it did have a strengthened transmission, a bigger clutch combined with the belt pulley, new internal expanding shoe brakes, wider rear tyres and a more comfortable seat. There was also a Series 2 Mk I version for farmers and the Mk II had a higher top speed for threshing contractors. Materials were still in short supply in 1947 when Marshall advertisements apologised to farmers because, 'A shortage of supplies means that the waiting period for delivery of a new tractor is longer than we would wish it to be'.

Engine rotation was changed to clockwise on the 40 belt hp Series 3 Field Marshall launched in 1949. The power take-off shaft was no longer offset and the tractor had a dual range six forward and two reverse speed gearbox. Engine performance was improved on the Series 3A, made between 1952 and 1956. Some Series 3A tractors had the familiar green livery but others left the Britannia Works with a new orange colour scheme and silver trim. The Series 3A was also the first Field Marshall with optional electric starting and a bolt-on hydraulic lift system made by the Ardrolic Engineering Co in Scotland.

The first 40 hp Fowler Mark VF crawlers, built at Leeds in 1948, were based on the Series 3 Field Marshall with a six forward and two reverse gearbox, controlled differential steering and power take-off. Fowlers had planned to call the new tractor the FD5 in line with

199. The Field Marshall had a single-cylinder two-stroke diesel engine, a large diameter clutch, independent brakes and a belt pulley.

200. Most Series 3 Field Marshall tractors had orange paintwork.

earlier FD crawler models but in the event it was changed to the VF with 'V' representing 5 in Roman numerals. An improved Mark VFA based on the Series 3A Field Marshall replaced the earlier model in 1952 and, like the Marshall wheeled tractors, most of them were painted orange.

201. A six-cylinder 70 hp Perkins engine provided the power for the Track Marshall 70 crawler.

The Track Marshall 50 with a 48 hp Perkins L4 engine and transmission system similar to the VFA appeared in 1956. A restyled bonnet and orange paint gave it a very different appearance and optional extras included a rear belt pulley, electric lighting and a weather cab. An improved Track Marshall 55 with a 55 hp Perkins 4-270 engine appeared in 1959. The Fowler side of the Marshall organisation at Leeds introduced the first of four Fowler Challenger crawlers in 1951. The Challenger I with a 50 hp twin-cylinder two-stroke

Marshall ED.5 engine had a six forward and two reverse speed gearbox. Driving controls included a gear lever, high/low range lever, two steering levers, a clutch pedal and a brake pedal that could be locked for parking. The Challenger 1 was only made for a short while but the 65, 80 and 150 hp Challenger 2, 3 and 4 crawlers remained in production until the factory closed in 1973.

Marshall designers were working on a replacement for the ageing Field Marshall in order to match the competition that had already used multi-cylinder diesel engines for several years. Their efforts resulted in the appearance of the pre-production Marshall MP4 with a 65 hp four-cylinder Meadows diesel engine at the 1954 Royal Smithfield Show. Production was due to start in 1955 but none were made until 1956 when the renamed Marshall MP6 with a six-cylinder Leyland 70 hp four-stroke, direct injection diesel engine and a six forward and two reverse speed gearbox

202. Many Marshall enthusiasts consider the 70 hp MP6 to be the last real Marshall wheeled tractor.

was produced. The new orange MP6 with silver trim was provisionally priced at £1,400, optional extras included a belt pulley, power take-off, cast-iron wheel weights and a 12-volt lighting set. However, it found little favour with British farmers and less than 200 were actually made. Only ten were sold on the home market and the last one was built in 1960.

The model number of the Track Marshall 55 indicated the horsepower of its Perkins L4 engine. The Track Marshall 70 launched in 1961 had a 70 hp six-cylinder Perkins power unit and a manual steering system. The 70C was the standard model but the 70H introduced in 1962 had hydraulic steering. In 1964 the orange colour scheme was changed to bright yellow. The slightly more powerful Track Marshall 56 and 75C replaced the 55 and 70 in 1965 and the new Track Marshall 90 was added the same year. Weather cabs and hydraulic linkage were optional and low ground pressure versions of the 56 and 75 hp tractors with wide tracks designated the Track Marshall 56W and 75W were also made.

When the Ward group bought Bristol Tractors in 1970 Marshall was renamed Marshall-Fowler Ltd, the Fowler works at Leeds was closed in 1973 and British Leyland Special Products Group bought Marshall-Fowler in 1975. The new owners used the Aveling-Marshall name for the 56, 75C and 90 tractors and later models until Lincolnshire farmer Charles Nickerson bought the company in 1979. He revived the original Track Marshall name and the existing Aveling-Marshall crawlers became the Track Marshall 100, 105, 120 and 140. The Track Marshall TM135 with a 136 hp six-cylinder Perkins engine, an air-conditioned cab, a single steering lever and running track with sealed lubrication was launched at the 1980 Royal Show. The 70 hp Britannia was added to the

Track Marshall range in 1982. Taking its name from the Gainsborough Works, the Britannia had a four-cylinder Perkins engine, hydraulic clutch and brake steering, power take-off and Marshall's own three-point linkage. Track Marshall, which was making the Britannia, TM110, TM120 and TM135 at Gainsborough, was the only British crawler tractor manufacturer when Charles Nickerson bought the Leyland wheeled tractor business in 1982.

The Marshall name reappeared on wheeled tractors after a gap of twenty-four years when the revived Marshall, Sons & Co Ltd introduced the harvest gold and black Marshall 502, 602/604, 702/704 and 802/804 tractors. Production continued at the old Leyland truck factory at Bathgate for a while and was later moved to Gainsborough. Marshall sales literature suggested that owners would no doubt agree that the four-wheel drive 804 was one of the finest of its type available at a price that offered excellent value with electro-static paintwork, quartz halogen ploughing lamps and radial tyres.

The Nickerson era proved short lived but Marshall Sons & Co Ltd introduced improved 62, 72 and 82 hp models and launched the new Marshall 115, 100 and

203. The Marshall 802 was the two-wheel drive version of the 80 hp Marshall 804.

904 XL wheeled tractors in time for the 1984 Royal Smithfield Show. There were ten different models at the show with the Field Marshall name and baton logo on the front grille of the angular-styled Marshall 100 and 115. The 103 hp Marshall 100 with a choice of two- or four-wheel drive and the four-wheel drive 115 hp Marshall 115, advertised as a brand new supremo for the Marshall army, had six-cylinder Leyland engines. Both had a twenty forward and nine reverse gearbox with an

204. The Marshall 100 with a turbocharged 100 hp Leyland engine was launched at the 1984 Royal Smithfield Show.

optional creep speed box, electronic linkage control hydraulics with remote switches on the rear mudguards and a flat floor Explorer cab. The turbocharged 92 hp 904XL was an improved version of the 804 with a fifteen forward and five reverse synchromesh transmission. The same fifteen forward and five reverse gearbox and a slightly longer wheelbase were the main changes for the improved two- and four-wheel drive 602/604, 702/704 and 802/804XL tractors.

Marshall Sons & Co Ltd went into receivership in 1985. Bentall-Simplex Industries rescued the wheeled tractor operation which, trading as Marshall Tractors Ltd, moved to Scunthorpe. The tracklayer side of the business changed hands yet again, this time it was bought by Herbert Flatters and became Track Marshall of Gainsborough Ltd.

205. The 35 hp Marshall 132 with a Duncan safety cab was launched in 1986.

With no available stand space at the 1986 Royal Smithfield Show Marshall Tractors introduced the compact two-wheel drive Marshall 132 to the farming public at a nearby hotel. The tractor with a Duncan cab and category I hydraulic linkage was assembled at Scunthorpe. It had a 35 hp engine based on a Perkins power unit and the transmission was made in Yugoslavia. The 184 and 264 four-wheel drive compact models with either a two-cylinder 18 or 26 hp Ruggerini or Lombardini air-cooled diesel engine and 'a light touch but firm tread' completed Marshall's light tractor range.

Production of Track Marshall crawlers including the new 155 hp TM155 continued at Britannia Works. Nine new Field Marshalls with 75, 85 and 95 hp Perkins engines and the baton logo on the radiator grille were announced in 1987. Made at Scunthorpe, the new Field Marshall 752/754X and 852/ 854X tractors had a nine forward and three reverse synchromesh gearbox while the XL version of these tractors and the four-wheel drive 954 XL had fifteen forward gears and five in reverse. The Marshall Explorer cab was retained but unlike earlier cabs it had a separate heating and ventilating system with an easily serviced dust filter.

206. The 155 hp Track Marshall 155 with five forward and reverse gears and category III hydraulic linkage cost £41,300 in 1988.

There was another change of ownership in 1987 when Tom Walkinshaw Racing (TWR) bought the Gainsborough business. However, sales did not come up to expectations and the Track Marshall range gradually disappeared but new heart was put into the Scunthorpe company in 1989 when an agreement was made with Steyr Daimler Puch to sell the Austrian-built Steyr tractors with the Marshall livery in the UK. Later that year, trading as Marshall-Daimler Ltd at Scunthorpe, a range of 64–150 hp two- and four-wheel drive Steyr D series tractors with harvest gold and black paintwork and the Field Marshall baton logo appeared on the UK market. Prices ranged

207. The two- and four-wheel drive Marshall D Series tractors were made by Steyr-Daimler-Puch.

from £18,395 for the 64 hp two-wheel drive D-642 to £45,800 for the four-wheel drive D-150 with a four-cylinder 150 hp engine under its gold-painted bonnet. Marshall-Daimler predicted the D series with a twelve forward and four reverse gearbox and the proven Explorer 2 control centre (cab) would take them to fifth place in the UK tractor sales league table within three years. Optional equipment for the 72 hp upwards D

208. A hand-operated hydraulic ram was used to tilt the cab when servicing the TM200 crawler tractor.

series included electronic load sensing hydraulics, D-matic thirty-six forward and twelve reverse change-on-the-move transmission and the Marshall-Daimler Informat computerised driver information system. Informat monitored engine performance, recommended the most efficient engine speed and gear ratio for maximum fuel economy and automatically controlled D-matic transmission gear selection. Originally marketed in Britain as the Steyr D series they were sold elsewhere in the world as Steyr tractors and with red and white paintwork.

The Marshall S series, also made by Steyr, was launched in 1990. The six-cylinder S-542/S-544 and S-624/S-644 two- and four-wheel drive tractors had 56 and 64 hp engines. The specification of the four-wheel drive 72 hp S-744 included a sixteen forward and eight reverse gearbox, live hydraulics and two-speed power take-off. However the relationship with Steyr was short lived and the company re-emerged as Marshall Tractors, now reduced to selling spare parts and re-built tractors from the Scunthorpe factory.

The 70 hp Britannia crawler was discontinued in the late 1980s when plans were in hand for a new rubber-tracked Track Marshall. The TM200, which made its debut at the 1990 Royal Show, was the first real competitor for the rubber-tracked Caterpillar Challenger and about 1,000 of the rubber-tracked American crawlers had already been sold worldwide since its launch in 1988. The first TM200 crawlers had a 200 hp turbocharged Cummins engine and sixteen forward and two reverse hydro power shaft transmission and closed centre category III hydraulics. A later version

with a 210 hp Cummins power unit had a two-speed hydrostatic transmission with a hydraulic motor to drive each track. The steering wheel operated a control valve that slowed the inner track when changing direction. The Australian-designed rubber tracks had lugs on the underside of the track, which engaged with the rear driving sprockets, and a wheeled tractor-type pivoting front axle was linked to a pneumatic suspension system. The TM200 was not a big seller, it was discontinued in 1994 and the Track Marshall TM155 was only made to order until production came to an end in 1996.

Martin Markam

Martin Markham at Stamford in Lincolnshire, manufacturers of a range of farm machinery including cultivators, fertiliser distributors and forage harvesters, made about 200 standard and de luxe Colt tractors between 1961 and 1970. Marketed by Colt Tractors the standard model had a 7 hp air-cooled Kohler engine with a recoil starter. A 12-volt electric system with a starter motor, dynamo and battery was an optional extra. Power was transmitted by a flat belt to a single-plate dry clutch, then by a shaft to a transfer box and roller chain drive to a three forward and one reverse gearbox. The metallic blue tractor with red wheels had a single foot brake with a parking latch; the rear wheel track was adjustable in 4 in steps from 32 in to 40 in and the front track from 29 in to 37 in. A gear pump with an output of 4 pints per minute supplied oil to the four-point hydraulic linkage ram cylinder and to an external ram connection for the Martin Markham front loader and tipping trailer. The Colt had three power take-off shafts, the one at the rear turned at 280 rpm while the front power shaft for use with a vee-belt pulley and a centre shaft for a flat belt pulley both ran at approximately half engine speed.

The de luxe Colt, similar to the standard model but with a 10 hp Kohler engine and electric starter, was launched at the 1965 Royal Smithfield Show. Features included a six forward and two reverse gearbox, four-point hydraulic linkage, two power take-off shafts, independent foot brakes that could be locked together when driving on the road, a hand brake and a padded seat. A sales leaflet explained that the 'light automotive steering, simple gear change and fingertip hydraulics make the Colt de luxe so easy to control that anyone can learn to drive the tractor in a matter of minutes'.

Maskell

Essex farmer John Maskell experimented with various types of motorised plough in the early 1920s and in 1924 he made a three-wheel motor plough with a Ford Model T car engine. After moving to Dorset in the early 1950s the family designed and built a small self-propelled tool frame. The Maskell self-propelled toolbar, launched at the 1961 Spring Sugar Beet demonstration, was made by H Maskell & Son at Wilstead near Bedford. It had a 14 hp Enfield 100 twin-cylinder air-cooled diesel engine, a three forward and one reverse gearbox and a differential reduction unit mounted on a hollow square section steel frame above the roller chain-driven rear wheels. The Maskell had independent steering brakes and the single or optional double front wheels were steered with a tiller handle. Hydraulic linkage and an on-board hydraulic motor, which could be attached at various points on the

209. Sales literature described the Martin Markham Colt as a 'small tractor with a great performance and the most practical and versatile machine in its class'.

210. An optional hydraulic motor, which could be mounted on the Maskell rowcrop tractor tool frame, was used to drive seeder units and other power-driven attachments.

tool frame, were optional extras. An 8½ gal/min engine-mounted pump supplied oil from a 15-gallon reservoir to the lift ram. The Maskell, which used less than a gallon of fuel per hour, had a minimum ground clearance of 17 inches and a top speed of 8 mph. Barfords of Belton in Lincolnshire, who were part of the Aveling-Barford Group, made a redesigned Barford-Maskell rowcrop tractor, also with a rear-mounted engine and transmission, in the mid-1960s. It had a high arched rectangular hollow steel tool frame and an engine-mounted hydraulic pump supplied oil to the double-acting ram used to raise and lower the tool frame into and out of work.

Massey Ferguson

Nearly 360,000 Ferguson TE20 tractors had been made at Banner Lane in Coventry when Massey-Harris and Ferguson amalgamated in 1953 but no immediate changes were made to either company's product range. The grey and gold Ferguson FE35 with a four-cylinder Standard petrol, vaporising oil, diesel or 29 hp lamp oil engine replaced the TE20 in 1956. The Massey Ferguson badge did not appear on Coventry-built tractors until the MF35 superseded the FE35 in 1957. The four-cylinder diesel engine was retained and the change to the red and grey MF35 colour scheme was the only difference between the two tractors. The petrol- and diesel-engined MF 35 tractors were rated at 37 hp and the vaporising oil model at 30 hp. The de luxe MF35 had a dual clutch with a live power take-off and hydraulic system.

The Mk I MF65 launched in 1957 was a big brother for the MF35. The Massey Ferguson publicity department celebrated the event by advertising that 'together the 35 and 65 tractors make 100 per cent Massey Ferguson farming available to all'. The 50½ hp MF65 with a four-cylinder Perkins diesel engine had many of the features found on the MF35 together with inboard disc brakes and epicyclic final drive reduction units. Optional extras included a diff-lock and power steering.

Massey Ferguson took over the lease for the Banner Lane factory from the Standard Motor Co and also bought Perkins Engines Ltd at Peterborough in 1959. The 39.9 hp MF35 tractor with a three-cylinder Perkins engine was launched the same year and tractor drivers who had struggled to start the earlier four-cylinder Standard diesel engine on a cold morning welcomed the change. In praise of the three-cylinder MF35 a sales brochure suggested it 'has a place on every farm and will give its owner extra energy to tackle fresh work and bring profit from every single acre'!

MF65 tractors were not at their best with trailed implements so Massey Ferguson introduced the Multi-Pull hitch in 1960. The predecessor to Pressure Control it had a heavy chain attached to a frame on the three-point linkage which was wrapped around a trailed implement drawbar. Partially raising the hydraulic lift arms transferred some of the implement weight on to the back of tractor and improved wheel grip.

211. The Mk I MF65 introduced in 1957 had a four-cylinder Perkins diesel engine.

The 56.8 hp Mk II MF65 with a diff-lock and optional road lights was introduced at the 1960 Royal Smithfield Show. When the National Institute of Agricultural Engineering tested the tractor at Silsoe the engine developed a maximum of 58.3 hp and Massey Ferguson lost no time in publicising the MF 65's extra power. An optional factory fitted twelve forward and four reverse speed Multi-Power gearbox was announced in 1962. The new Multi-Power change-on-the-move system cost an extra £70. It had a hydraulic clutch controlled by a switch on the instrument panel that gave a 30 per cent speed increase or decrease in each gear. Optional equipment for the MF35 was increased in 1962 to include a diff-lock and power-adjusted variable track (PAVT) rear wheels but Multi-Power was not available until the 44½ hp MF35X was launched at the 1962 Royal Smithfield Show. Although most tractors were sold with a diesel engine in the early 1960s it was still possible to buy the MF35 with a petrol or vaporising oil engine but the 35X was only available with a diesel engine. The 77 hp diesel-engined MF Super 90, made by Massey Ferguson in America, was another new model on their stand at the 1962 Royal Smithfield Show. Priced at £1,980 the Super 90 had an eight forward and two reverse gearbox, power-adjusted rear wheels, disc brakes and power steering. The pump for the hydraulic system was located in the gearbox and it used transmission oil cooled by a radiator at the front of the tractor. The Super 90 had a lower link sensing draft control system with a heavy-duty double-acting assistor spring. A switch on the instrument panel was used to adjust the level of response made by the hydraulic system to changes in the load on the lower links.

The 'Red Giant' 100 series tractors with square-shaped bonnets and lights built into the radiator grille replaced the MF35 and MF65 in December 1964. Flat top wings were used for the 135, 165 and 175 with a flexible cladding weather cab but the MF130 Economy and the cab-less MF135 had round mudguards.

The 30 hp French-built MF130 Economy model had a Perkins four-cylinder indirect injection diesel engine. The standard specification included an eight forward and two reverse gearbox with synchromesh on third to fourth and seventh to eighth, disc brakes, diff-lock and a mechanical linkage lock to hold implements in the raised position for transport. The de luxe 130 had a dual

212. A steel-framed fibreglass weather cab with a spring suspension seat was an optional extra for the four-cylinder 66 hp MF175.

clutch and a central power take-off shaft for mid-mounted implements. However, the MF130 was not popular and was discontinued in 1972.

The Perkins-engined 135 and 165 were very similar to the MF35 and Mk II 65 and the slightly more powerful MF35X engine used for the MF135 was rated at 45½ hp. The MF175 with a 66 hp four-cylinder Perkins-engine, PAVT rear wheels with heavy cast-iron centres and power steering was in other respects very similar to the MF165. Pressure Control, which replaced the earlier Multi-Pull hitch system, was optional on the MF165 and 175 but within a few months it was standard. Pressure Control could be used to transfer up to a ton of trailed implement weight onto the back of the tractor but without a suitably strengthened implement drawbar it was liable to bend and sometimes did. Optional extras for the MF135, 165 and 175 included Multi-Power transmission, hydraulic spool valves, a weather cab, a foot throttle, a spring suspension seat and a cigarette lighter.

Various improvements were made to the Red Giants during their twelve-year production run. The MF165 engine power was increased to 60 hp and the 66 hp MF175 became the 72½ hp MF178 in 1968. Other changes in the early 1970s included the addition of dry element air cleaners, oil-cooled brakes, improved hydraulics and an independent power take-off. Flat top

213. Made in Detroit, the MF1100 was introduced to UK farmers at the 1967 Royal Smithfield Show.

wings were standard across the range from 1970 when, in order to meet government regulations, all Massey Ferguson tractors sold in the UK had a safety cab.

The 96 hp Massey Ferguson 1100 was described as 'the biggest Red Giant of them all'. Made in Detroit in 1965, it was the most powerful Massey Ferguson tractor available in the UK when it was introduced to British farmers at the 1967 Royal Smithfield Show. The 1100 had a six-cylinder Perkins engine, a dual clutch, a twelve forward and four reverse Multi-Power transmission and hydrostatic steering. Other features included a two-speed independent power take-off, twin interconnected fuel tanks situated under the foot plate on each side of the tractor, PAVT rear wheels and a safety-start mechanism in the clutch pedal linkage. Two hydraulic pumps with an oil cooler provided oil pressure for the three-point linkage, auxiliary service rams, power steering, brakes, power take-off clutch and hydraulically suspended seat. Pressure Control was standard and the 1100 was also one of the first tractors with the facility to mix the use of the position and draft control hydraulics. The 90 hp two-wheel drive French built MF1080 launched in the UK in 1969 was similar to the earlier American MF180. The specification included a twelve forward and four reverse Multi-Power transmission, hydrostatic steering with an adjustable steering column, dry disc brakes, cast-iron wheel centres, PAVT rear

wheels and an optional 1,000 rpm power take-off shaft. A package of seventeen changes for the 92 hp Mk II 1080 included a much quieter cab, inboard oil-cooled brakes and the 1,000 rpm power take-off shaft was standard. Some manufacturers provided a pair of ear defenders to overcome the problem of the high noise level in safety cabs. This prompted MF to improve the sound insulation on the Mk II 1080 cab and an optional sound deadening kit costing £16 was introduced for the earlier Mk I cab.

Turbochargers, advertised as increasing engine power by about 20 per cent, and front-wheel drive conversions were becoming popular in the late 1960s. Robert Eden Ltd at North Audley Street in London imported Selene four-wheel drive kits from Italy for the Massey Ferguson FE35, MF35 and MF65 in the mid-1950s and ten years later, as Four Wheel Traction Ltd, sold similar conversion kits for the 100 series tractors. The mid-1950s Selene conversion kits had a transfer box attached to the power take-off shaft, which was used in ground speed, and a forward-running shaft transmitted power to a differential housing on the front axle. The power take-off shaft was extended through the transfer box for use with power-driven implements but it was necessary to disengage the drive to the front wheels and reset the power shaft speed at 540 rpm. Similar front-wheel drive kits were made for the MF135, 165 and 175 until the

214. The articulated 105 hp Massey Ferguson 1200 – first made in America in 1970 – appeared in the UK in 1971.

mid-1960s when Robert Eden made a strengthened Eden Manuel conversion kit under licence for MF tractors. The new design with the transfer box sandwiched between the gearbox and back end meant that the power take-off could be used at 540 rpm. Four Wheel Traction, established by Robert Eden in the late 1960s, made front-wheel drive conversion kits with either small or equal-sized front wheels for MF tractors until the late 1970s when Massey Ferguson introduced their own four-wheel drive tractors.

Eight forward speeds and longer wheelbases were among the features of the new high-specification MF148, 168 and 188 tractors introduced in 1971. The six-speed 135 and 165 were retained but the 178 was replaced by the MF185. Engine power was on its inevitable upward trend. The MF135 and 148 were rated at 47 and 49 hp respectively, the 165 and 168 had 62 and 69 hp under their bonnets and 75 hp engines were used for the 185 and 188.

Massey Ferguson launched the 150 hp articulated four-wheel drive MF1500 and 105 hp MF1200 in America in 1970. The four-wheel drive Massey Ferguson 1200, which made its UK debut at the 1971 Royal Smithfield Show was by far the biggest MF tractor yet seen in the UK. Made at Manchester from 1972 the six-cylinder MF1200 with equal-sized wheels had a twelve forward and four reverse speed Multi-Power transmission and the centre-pivot articulated steering with double-acting hydraulic rams provided a 12 ft

turning circle. Opico and some other companies offered turbocharger kits for existing MF185, 188 and 1200 tractors in the early 1970s. The 1,000 rpm power take-off shaft on the turbocharged 1200 developed 106 hp compared with 87 hp on the naturally aspirated model.

The 88 hp MF595, which superseded the 1080 in 1974, was the first of the new square-fronted Massey Ferguson 500 series designed to replace the popular but ageing 100 series tractors. The two- and four-wheel drive 595 had a single-plate dry clutch and the usual twelve forward and four reverse Multi-Power transmission with a top speed of 20.4 mph. The drive shaft to the front axle was under the engine and a switch was used to engage or disengage front-wheel drive while on the move. The flat floor cab, isolated from the chassis on anti-vibration rubber mountings, had full instrumentation, large areas of glass and a spring suspension seat. Sales literature described the MF595 as a tractor with 'a super comfort, pressurised cab with heating and ventilation systems'.

The two-wheel drive MF1135 with a turbocharged six-cylinder 135 hp Perkins engine and the MF1155 with a 155 hp Perkins V8 engine, launched at the 1975 Royal Smithfield Show, had a similar specification to the earlier 1080. The 'roomy super comfort cab' which provided 'a dust-free, low noise level working environment' could be tipped backwards to service the gearbox and rear axle. The 1155 was made until 1977 and the 1135 was discontinued in 1979.

The MF550, 565, 575 and 590 with 47, 60, 66 and 75 hp Perkins engines made at Banner Lane joined the earlier French-built MF595 in 1976. The new models with an integral single door cab mounted on rubber blocks had increased hydraulic pump capacity and a second gear pump supplied oil to the power steering and other auxiliary circuits and they were the first MF tractors with an alternator charging system. Optional equipment included an eight-speed manual or twelve-speed Multi-Power transmission and an independent power take-off or a live power shaft with a dual clutch. Prices ranged from £4,100 for the basic MF550 to £5,578 for the top of the range MF590 with Multi-Power transmission. Optional two- or four-wheel drive for the 570 and 590 were added in 1978.

215. The cab on the 135 hp MF1135 was tipped back when servicing the gearbox and rear axle.

Previously launched in America in 1972 the most powerful articulated four-wheel drive Massey Ferguson 1505 tractor to be introduced in the UK appeared at the 1975 Royal Smithfield Show. The MF1505 with a 180 hp direct injection Caterpillar V8 diesel engine, an air-conditioned cab on rubber mountings, category III hydraulic linkage and a 1,000 rpm power take-off rated at 160 hp was discontinued in 1978.

The MF 200 series replaced the standard 135 tractor in 1979 but the orchard, vineyard and crawler versions of the tractor were made until 1982. The French-built 135 vineyard and the

216. The MF1505 had a 180 hp Caterpillar V8 diesel engine.

QD orchard models had a 45 hp Perkins AD 3.152 diesel engine, a quick-detach (QD) safety cab was standard on the 135QD orchard model and the vineyard tractor had a quick-detach safety frame. A special three-point linkage (category V) with cranked lower links for the vineyard model was introduced in 1976. The 135 engine was de-rated to 41 hp on the 134C crawler tractor built by Landini in Italy who also made the 47 and 61 hp Perkins-engined MF154C and MF174C tracklayers. The specification included a dual clutch with separate hand levers for the transmission and power take-off, an eight forward and four reverse gearbox, clutch and brake steering, and live hydraulics with an engine-mounted pump. A dealer-fitted British-built cab made by Westlode Engineering at Spalding, with an optional heater, was available for the MF154C and 174C.

The 45 hp MF240 and the 60 hp MF265, modified versions of the earlier 100 series tractors, were the first of the new MF200 series tractors. These relatively basic tractors had an

217. The first Massey Ferguson 200 series tractors were made at Banner Lane in 1979.

eight-speed transmission, drum brakes, manual steering and a quick-detach cab. Optional extras included Multi-Power, power-assisted steering, an automatic pick-up hitch and pressure control. The cab was equipped with lifting eyes and could be removed in less than ten minutes when the tractor was used in low buildings. The 200 series was extended in 1981 with the three-cylinder-engined 33 hp MF230 and four-cylinder 75 hp MF290. The MF230 was a basic two-wheel drive tractor with an eight-speed gearbox and safety frame. The MF290 with the choice of an eight forward and two reverse, a twelve forward and four reverse synchromesh gearbox or a twelve-speed Multi-Power transmission had a detachable Duncan safety cab. Multi-Power transmission was an option for the 47 hp three-cylinder MF250 with an eight-speed manual gearbox introduced in 1982. It was also the first 200 series tractor with power steering and oil-cooled brakes. Two- and four-wheel drive versions of the 66 hp MF275 with a low-profile cab and 88 hp MF298 with twelve forward and four reverse gears were added in 1985. The higher specification MF298 had hydrostatic steering, hydraulic wet disc brakes and a two-speed independent power take-off. A sales drive at the 1985 Royal Smithfield Show offered 500 gallons of free fuel

with every top of the range 200 series tractor bought at Earls Court.

The Massey Ferguson 2000, 4000 and 600 series tractors were launched in the late 1970s and early 1980s. The French-built two- and four-wheel drive MF2640 and MF2680 rated at 104 and 120 hp were introduced in 1979. The 130 hp MF2720 and the 93 hp MF2620 were added in 1981 when the power rating of the earlier models was increased to 110 and 130 hp respectively. The 2000 series had a sixteen forward and twelve reverse Speedshift transmission with shuttle reverse, inboard disc brakes and a hydraulically engaged diff-lock. The lower link sensing hydraulic system had external lift rams and the driving controls were located on the right-hand side of a spring suspension seat in a high-visibility cab. The 1200 was discontinued in 1980 but not before the articulated 112 hp MF1250 with an improved transmission and hydraulic system had been added to the Massey Ferguson range.

Following an evaluation exercise in eastern England the articulated four-wheel drive Massey Ferguson 4840 was introduced to the British market in 1980, coincidentally exactly fifty years since the first four-wheel drive Massey-Harris General-Purpose tractors were made in 1930. The MF4840, introduced two years earlier in North America, had a 260 hp Cummins V8 engine, an oil-cooled multi-disc clutch and there were three changes on the move in each of the six gear ratios in the eighteen forward and six reverse gearbox. The 4840 was also the first tractor in the world to have an electronically controlled hydraulic linkage with a series of induction coils which sensed changes in the load on the tractor and transmitted signals to a

218. The 260 hp Massey Ferguson 4840 with a Cummins V8 engine was introduced to UK farmers in 1982.

solenoid unit linked to the hydraulic control valve. The driver enjoyed the comfort of an upholstered, swiveling seat in an air-conditioned cab, which he had to leave occasionally to fill the tractor's 160-gallon fuel tank. North American farmers were also able to buy the 225 hp MF4800 and 315 hp MF4880 in the early 1980s.

The two- and four-wheel drive, four-cylinder MF675, 690 and 698 launched in 1981 and rated at 66, 77 and 88 hp were the first MF600 series tractors. They had a twelve forward and four reverse synchromesh gearbox or optional Multi-Power transmission, hydraulic brakes, hydrostatic steering and a flat floor cab. The six-cylinder 98 hp MF699 was added in 1984 when the 90 hp MF698T with a wastegate turbocharger replaced the MF698. Wastegate turbochargers give a high power boost at low engine speeds and a spring-loaded dump valve (or wastegate) prevents an excessive power surge when the engine approaches its maximum speed.

At the lower end of the power scale Massey Ferguson still made special-purpose tractors which began with the TE20 vineyard model in the early 1950s and continued with narrow versions of the FE35, MF35 and MF135. The Italian-built MF145 vineyard and MF158F fruit tractor replaced the long-serving specialist 135 models in 1982.

219. Launched in 1981 the Massey Ferguson 698T and the other 600 series tractors had the usual red bonnet and cab but the rest of the tractor was painted with a new shade of MF charcoal grey.

The 48 in wide MF145 had a 45 hp three-cylinder Perkins engine and the 59 in wide MF 158 had a four-cylinder 55 hp Perkins power unit. Both tractors, with the familiar Massey Ferguson 135 square-shaped bonnet, had an eight-speed synchromesh gearbox and a 'zero-leak' hydraulic ram acted as a lock for the three-point linkage when the engine was stopped.

Massey Ferguson decided to test customer reaction to a range of small Japanese tractors at the 1983 Royal Show and a favourable response resulted in the launch of the MF1010, 1020 and 1030 compact models in 1984. The two- and four-wheel drive 16, 21 and 27 hp diesel tractors cost from £3,790 for the two-wheel drive MF1010 to £5,650 for the four-wheel drive MF1030 with more horsepower under its bonnet than the original Ferguson 20.

220. Introduced in 1983 the Massey Ferguson 294 C was the largest of five 45–73 hp crawler tractors made by Landini in Italy.

The MF2005 series replaced the 2000 tractors in 1985. Improvements including a new two-door cab with a roof hatch, increased hydraulic capacity and engine power were the result of a questionnaire sent to 4,000 farmers. However, electronic linkage control (elc) with sensor pins to measure draft forces in the lower links was the most significant new feature on the MF2645, 2685 and 2725 with six-cylinder engines rated at 110, 130 and 147 hp.

The MF3000 series with the new computerised Autotronic and Datatronic systems introduced in 1986 ranged from the four-cylinder 68 hp MF3050 to the 107 hp MF3090 with a six-cylinder engine. The standard specification included a hydraulic self-adjusting clutch, a sixteen forward and reverse synchromesh gearbox with two 'H' gate gear levers and electronic linkage control hydraulics. Dubbed 'the intelligent tractors with real muscle and brainpower' sales literature for the 3000 series explained that the tractors could be programmed to select the most cost-effective way of working to give peak performance and productivity with maximum economy. The Autotronic computer disengaged the diff-lock when the hydraulic linkage was raised and re-engaged it when the implement was lowered and disengaged the power take-off if the implement was blocked or ran too fast. Autotronic also engaged four-wheel drive when the driver applied the brakes and disengaged four-wheel drive at speeds in excess of 9 mph. The more sophisticated and expensive Datatronic system also controlled wheelslip and provided information on the tractor's performance to help the driver achieve maximum efficiency.

Seven 47–97 hp MF300 series tractors with improved transmissions, hydraulics, steering and cabs replaced the top five models in the 200 series in 1986 but the MF230 and 240 were retained to provide relatively basic tractors at the lower end of the power range. There were five gearbox options for the two- and four-wheel drive 300 series and the four-cylinder 90 hp MF399 was the first Banner Lane tractor with a turbocharged engine. The 58 hp MF360 replaced the 355 in 1987 and the six-cylinder 90 hp turbocharged 390T in two- or four-wheel format was added in 1989. An optional low-profile cab was available for all 300 series tractors except the MF399 in 1988 when a new silver-painted Hi-Line flat floor cab was introduced for the 78–110 hp 300 series tractors. The earlier temptation of 500 gallons of free fuel for farmers buying selected models in the 200 series tractors was repeated in metric form in 1989 when the bait of 2,000 litres of free diesel was offered to farmers buying specified MF300 and MF3000 models.

The two- and four-wheel drive 59 hp MF362 with eight forward and two reverse gears or optional eight-speed shuttle transmission, hydrostatic steering and hydraulically engaged power take-off was added in 1990 and there were more changes to the 300 series in the early 1990s. A twelve-speed synchromesh reverse shuttle transmission became standard on Hi-Line cabbed models from the 71 hp MF375 up, a new 104 hp six-cylinder Perkins Quadram engine was used on the MF399 and there was a new optional twelve-speed transmission for the larger four-wheel drive tractors. Some low-specification models of the 300 series tractors were made at Banner Lane in the early 1990s for overseas markets.

The more powerful MF3600 series tractors, added to the MF3000 series in 1987, were advertised as the 'thinking tractors with power to spare'. The 113, 133 and 150 hp MF3610, 3630 and 3650 Datatronic and Autotronic models had a sixteen-speed synchromesh reverse shuttle gearbox, improved lower link sensing, elc hydraulics and oil-cooled disc brakes. Changes came thick and fast in the early 1990s. A key-operated engine stop control was

221. Improvements for the 147 hp Massey Ferguson 2725, which replaced the 2720 in 1985, included new two-door cabs and electronic hydraulic linkage control.

222. Features of the mid-1980s French-built 93 hp Massey Ferguson 3070 included automatic four-wheel drive, which was engaged when braking. Drive to the front wheels was disengaged at speeds over 8½ mph.

introduced in 1990 for some MF3000/3600 tractors which were also the first to have the facility to download performance and field data from the in-cab Datatronic monitoring system to the farm computer. The MF3600 models with a sixteen-speed reverse shuttle gearbox, included the 142 hp 3645 and 155 hp 3655 with Perkins engines but 170, 180 and 190 hp Valmet engines were used for the MF3670, 3680 and 3690 tractors.

The Massey Ferguson optional Active Transport Control system designed to improve driver safety and comfort on the 3000, 3100 and 3600 series was introduced in 1991. ATC used a nitrogen accumulator in the hydraulic circuit to smooth out shock loads caused by implement bounce during transport.

223. *The 6280 with a 132 hp Perkins engine was one of the eight Massey Ferguson 6200 series tractors launched in 2000.*

The Dynashift transmission and the MF3065 HV sloping bonnet line achieved by moving the radiator header tank to the rear of the engine compartment were Massey Ferguson's innovations for 1992. The thirty-two forward and reverse speed Dynashift system with twenty-four clutchless powershift changes was standard on the 132–190 hp MF3600 series.

The turbocharged 120 hp MF3120 replaced the MF3115 in 1992 and the Perkins Dual Zone torque engine for the MF3100 and 3600 series appeared the same year. Precise control of the fuel injection system on the Dual Zone engines provided separate torque ranges for light and heavy work. This enabled the engine to respond quickly to an increasing load when ploughing or provide an economical power output for lighter work. Massey Ferguson also supplied Iseki with four models from the MF3000 and 3100 series with 88–125 hp engines and crystal blue paintwork in the early 1990s.

The Allis Gleaner Corporation (AGCO) formed in 1990 at Atlanta in Georgia, the distributors for MF tractors and machinery in North America, bought Massey Ferguson from the Varity Corporation in 1994.

The acquisition added the Massey Ferguson product range to their existing brands including Allis Gleaner, White and Hesston and by 1995 AGCO was marketing about seventy different models of tractor in the 13–425 hp bracket. They included 40–215 hp AGCO-Allis models, White tractors with 60–215 hp engines, articulated four-wheel drive 350 and 425 hp AGCOStar models and the full Massey Ferguson range. AGCO was also selling twelve specialist SAME tractors and a range of fifty-one Landini models in North America. The Varity Corporation retained the Perkins engine business and a management team bought the MF industrial equipment division which, trading as Fermec International, marketed Massey Ferguson-designed industrial equipment until the business was sold to Case IH.

The MF6000 and MF8000 series tractors replaced the 80–190 hp MF3000 and MF3600 series in 1995. The smallest MF6100 was equivalent to the earlier 80 hp MF3060 and the flagship MF8160 had a 200 hp six-cylinder turbocharged Perkins 1000 series engine. The specification included a multi-plate oil-cooled clutch, a thirty-two forward and reverse Dynashift pressure lubricated transmission, digital electronic hydraulic linkage control and a new generation of the MF Autotronic and Datatronic management systems.

Massey Ferguson made the three millionth tractor, a 66 hp MF375E, at Banner Lane on 15 October 1996. Tractors were also being made at Banner Lane for AGCO-Allis, Iseki, Landini, Valmet and White when the new MF4200 series was launched in 1997. The new Coventry-built 4200 series was introduced to farmers in thirty European countries in a four-minute spot on satellite television and to customers attending a breakfast-time launch at their local dealership. Features of the 52–110 hp MF4200 range included the choice of a Lo-Profile,

224. The 2005 Massey Ferguson 8480 with a common rail four-valve Sisu engine had an electronic engine management system.

Standard and Hi-Visibility cab, a raised central drive shaft to the front wheels on four-wheel drive models and automatic engagement of the front diff-lock when the rear diff-lock was used.

The most powerful Massey Ferguson tractor yet made in Europe was introduced to British farmers at the 1998 Royal Smithfield Show. The new flagship MF8180 with a 260 hp turbocharged and intercooled six-cylinder Valmet engine and an eighteen-speed powershift transmission developed 225 hp at the power take-off.

The 2001 Massey Ferguson price list included thirty-one tractors in the 54–260 hp bracket. They included the 3300 and 4200 series in two- and four-wheel drive format and some of the 4200 series were available with a low-profile cabs and a High Visibility sloping bonnet. The four-wheel drive MF6200 and MF8200 series, with the exception of the four-cylinder 85 hp 6245 and 95 hp 6255, had six-cylinder engines. The 4200 series was improved in 2001 with a PowerShuttle gearbox providing clutchless shifting between forward and reverse, electronic linkage control, wider and quieter cabs and optional front linkage and power take-off. New features of the 6200 and 8200 series included the Massey Ferguson QuadLink front suspension system, the

Autodrive automatic gear change system for the Dynashift transmission and an optional Spool Valve Management System which gave more accurate control of mounted and trailed implements. The Banner Lane factory closed in 2002 after more than 3,300,000 tractors had been made at Coventry since the first Ferguson TE20 tractor came off the production line in 1946.

Optional Cultivation, Transport and Land Technic packages together with a Comfort or Comfort Plus package for the cab were available for the new MF6400 and MF7400 series launched in 2003. The Cultivation package included engine speed control, engine power boost, Autodrive, load-sensing hydraulics and four extra work lamps while engine speed control, power boost, Autodrive and QuadLink front suspension were included in the Transport package. A super de luxe air suspension seat, passenger seat, rear-screen wash/wipe, electrically adjusted wing mirrors and automatic air conditioning were included in the Comfort Plus package. Features of the new 120–185 MF7400 range included Dyna-VT (CVT) stepless transmission with a 50 kph road speed, front axle and cab suspension and closed centre load-sensing hydraulics.

The launch of the 8400 series of Massey Ferguson

tractors in 2004 completed a two-year programme of model replacement which had seen the introduction of the 2400, 3400, 5400, 6400, 7400 and 8400 series tractors to the farming public. The 215–290 hp 8400 series had Dyna-VT continuously variable transmission, improved elc hydraulics and the latest Datatronic III terminal for automatic headland control, spool valve management and work displays at the driver's fingertips.

Massey-Harris

The Massey Manufacturing Co and A Harris & Son merged in 1891 to form Massey-Harris. Daniel Massey made his first farm implements including cutter bar mowers, reapers, threshers and other farm implements in his Ontario workshop in 1847. His eventual partner Alanson Harris established a farm machinery business ten years later at Beamsville, also in Canada. Massey-Harris sold their first tractors in 1917 when they introduced the American-built Bull tractor to Canadian farmers. When the Bull Tractor Co failed in 1919 Massey-Harris sold the M-H1, M-H2 and M-H3 tractors based on models made by Dent and Henry Parrett in Chicago. The M-H1 was sold in the UK as the Clydesdale but the Parrett designs were soon outdated and production ceased in 1923. No more Massey-Harris tractors were sold until 1928 when they bought the JI Case Plow Co factory at Racine, Wisconsin and in the

same year sold the JI Case Plow Co name to the JI Case Threshing Machine Co. Massey-Harris marketed the Wallis 20-30 tractor in America from 1928 and then added the smaller Wallis 12-20 in 1929. The Massey-Harris 25 replaced the Wallis 20-30 in 1931.

Massey-Harris opened factories in France and Germany between the two world wars and in 1930 they acquired a major interest in HV McKay at Sunshine near Melbourne in Australia. Massey-Harris tractors on UK farms at the time included the 12/20 and 20/30 with four-cylinder paraffin engines, three forward gears and one reverse and a belt pulley. A new red colour scheme replaced the earlier green paintwork on the Massey-Harris-badged Wallis tractors in the early 1930s when other Massey-Harris models included the 15-22, the Pacemaker and the Challenger followed by the Massey-Harris 80, 100 and 200 series tractors. The four-wheel drive Massey-Harris 15-22 or General Purpose model, made between 1930 and 1938 had a four-cylinder Hercules engine and a three forward and one reverse gearbox with a top speed of 4 mph. The Hercules side-valve paraffin engine, which developed 25 belt hp and 16 hp at the drawbar was used until 1936 when a slightly more powerful overhead valve Hercules engine replaced it. The design of the final reduction gears in the rear axle and its four large equal-sized wheels gave the General Purpose tractor a ground clearance of 30 in

225. There was a choice of steel wheels or pneumatic tyres for the Massey-Harris Pacemaker.

226. The Massey-Harris 744D was made at Kilmarnock between 1949 and 1954.

and the steel or optional pneumatic tyred wheels could be adjusted to give a track width of 48–76 in.

The 16–27 hp Massey-Harris Pacemaker introduced in 1936 had a four-cylinder water-cooled, overhead-valve paraffin engine with a tubular radiator, a 16 in cooling fan, centrifugal water pump and a mix of splash and force-feed lubrication. A farming magazine commented that the four-speed gearbox with a top speed of 8½ mph was 'very useful when moving the tractor from field to field'. The Pacemaker with power take-off, belt pulley and spade lug wheels cost £260 when 'delivered to the nearest UK railway station'. Dunlop low-pressure pneumatic tyres instead of steel wheels added £45 to the price.

Introduced in 1936 the Massey-Harris Challenger tricycle-wheeled rowcrop tractor was an improved version of the earlier 12-20 with adjustable track rear wheels on splined axles. The three-plough Challenger with a water-cooled four-cylinder overhead-valve engine, four forward gears and one reverse developed 16 drawbar hp and 27 hp at the power take-off.

Standard, rowcrop and tricycle-wheeled versions of the 39–45 hp Massey-Harris 44 and the 52–59 hp Massey-Harris 55 with the choice of a petrol or diesel engine were introduced to Canadian farmers in 1946. The 'one-plow' Massey-Harris Pony was added the UK range in 1947 to compete with the Allis-Chalmers Model B, Farmall A and Ferguson 20 tractors. Small-scale production of Massey-Harris mowers and hay machinery began at Trafford Park, Manchester in 1945. The company also sold the Canadian-built 44 and 55 tractors in the UK until the first Massey-Harris 744PD were made at Manchester in 1948. The prefix figure 7 indicated British manufacture. The 46 hp 744PD with a six-cylinder Perkins P6 engine, electric starter and Ki-gass cold starting system cost £854.10s ex-works. The specification included a five forward and one reverse gearbox, independent brakes, belt pulley, power take-off and wheel weights. Category II hydraulic linkage with the pump driven from the engine crankshaft was an optional extra. The tractor became the 744D when production moved to Kilmarnock in 1949 and about 200 of these tractors, including some Hi-Arch high clearance rowcrop models, were made each month. The two-part cast-iron chassis used for the M-H 44 tractor made in Canada was also shipped to the UK for the 744D. One part of the chassis supported the Perkins P6 engine and the other section carried the transmission. The five forward and two reverse gearbox had a top speed of 12 mph and the final drive included a spiral bevel crown wheel, spur tooth reduction gears and shoe brakes on the differential shafts.

The Massey-Harris 745 with a 44.6 hp four-cylinder L4 Perkins diesel engine and five forward and two reverse speed transmission replaced the 744D in 1954. The standard model, which cost £699, had a Hardy Spicer coupling from the clutch to the gearbox to facilitate clutch repairs. The rear wheels were fixed on the standard tractor but track width was adjusted on the rowcrop model by sliding the rear wheels along on their axle shafts and reversing the wheel centres. Optional equipment included the new 'instant action' hydraulic three-point linkage with the pump driven by the engine timing gears, twin front wheels, a Hi-Arch adjustable front rowcrop axle, field lights and a special 'velvet ride' seat. The 'Velvet Ride' seat in lieu of the standard spring leaf version and an hour meter both cost an extra £3.10s and the optional power take-off shaft was £10. A sales leaflet described the 745 as 'a genuine all-purpose four-furrow tractor, strong and versatile, with plenty of power for the heaviest work but equally efficient and economic for the smaller jobs around the farm'.

Following the amalgamation of Massey-Harris and Ferguson in 1953 the M-H745 was the last Massey-Harris tractor but there was one more production change at the Kilmarnock factory in 1957 when the M-H44 tractor was discontinued in Canada. As the 44 chassis castings were no longer available a steel-framed chassis was made at Kilmarnock for the modified and rather dated Massey-Harris 745S. The last of these tractors was built in 1958 when Massey-Harris-Ferguson became Massey Ferguson.

Matbro

The Matthew brothers started a general engineering business in Surrey in the late 1940s to manufacture Matbro forklift trucks and mechanical handling equipment. Matbro entered the agricultural market in 1962 with a modified four-wheel drive articulated tractor called the Matbro Mastiff 6/100MT which cost about £2,500.

The Mastiff had a six-cylinder 100 hp Ford diesel engine and two Fordson Super Major transmissions and rear axle units linked by a central articulating pivot. The engine was located above the front transmission unit and separate drive shafts transmitted power to both axles which were locked in permanent four-wheel drive. The Mastiff 6/100 had a diff-lock on both axles and a high capacity engine-mounted hydraulic pump provided oil pressure for the double-acting steering rams and the Super Major hydraulic linkage.

About twenty Matbro Mastiff 6/100 MT tractors were made before the articulated Mk II Mastiff with a

227. The Matbro Mastiff 6/100 was made between 1962 and 1967 at Horley in Surrey.

228. This articulated Mk II Matbro Mastiff was demonstrated to British farmers in 1967 but it failed to sell and spent its working days at Rotterdam docks.

128 hp Ford industrial engine appeared at several tractor demonstrations in 1967. The Mk II tractor with a Ford 5000 gearbox to transmit drive through a transfer box to two opposing Ford 5000 rear axle units, centre-pivot steering and hydraulic lift was priced at £3,495 but it failed to sell. The design concept reappeared in 1972 when Massey Ferguson launched the articulated four-wheel drive MF1200. Although Matbro enjoyed little success with tractors they were still manufacturing mechanical handling equipment in 2001.

McCormick

Cyrus Hall McCormick, who made a reaping machine in 1831, opened a factory at Chicago to manufacture his invention in 1847. The McCormick Harvesting Machine Co was incorporated in 1879 and in 1902 it merged with Deering and three other companies to form International Harvester. The McCormick name was used on tractors and farm machinery until Tenneco acquired the agricultural machinery division of International Harvester in 1984.

A condition of the Case IH/New Holland merger required the buyers to sell the International Harvester business. The Italian ARGO Group, which owned Landini, Laverda, Pegoraro and Valpadana, were the eventual purchasers of the Doncaster factory together with the McCormick name and the manufacturing rights for the Case IH CX and MC-X series tractors. Having been dormant since 1984 the McCormick name was revived in 2001 when McCormick Tractors International Ltd introduced the rebranded McCormick 73–102 hp two- and four-wheel drive CX and the four-wheel drive 90 and 102 hp MC tractors at Doncaster. Optional equipment for the CX and MC tractors included a de luxe or new low-profile air-conditioned cab and a more robust front linkage with double-acting lift rams.

McCormick Tractors added five new six-cylinder tractors, ranging from the 118 hp MTX110 to the 176 hp MTX175 with turbocharged Perkins engines, in 2002. There was a choice of a sixteen forward and twelve reverse powershift or a thirty-two forward and twenty-four reverse transmission with creep speed gearbox for the MTX175. Optional extras included front suspension, electronic power take-off control and a headland management system. The CX50 and CX60 with 53 and 64 hp engines were added later in the year. The 195 hp MTX 200 with a six-cylinder, twenty-four valve Cummins engine with full electronic fuel management was added to the existing five-model MTX range at the 2002 Royal Smithfield Show. The new model had a sixteen forward and twelve reverse powershift transmission with an optional push-button Speed Sequencer range selector.

The 2003 McCormick price list included the 54–93 hp F series, the 73–102 hp CX, the four- and six-cylinder 90–132 hp turbocharged MC tractors and the 118–195 hp six-cylinder MTX range. There was a choice of a sixteen forward and twelve reverse or a thirty-two forward and twenty-four reverse powershift transmission for the MC range and the XtraShift transmission was optional for F series and CX models. XtraShift was a three-speed powershift power shuttle transmission with wet multi-disc clutches which enabled the driver to change from forward to reverse.

McCormick International announced six new 200 hp-plus ZTX tractors with twenty-four valve Cummins engines, an eighteen-speed powershift transmission and a fifty-five degree steering angle and an updated MTX range in 2004. Optional extras for the more powerful MTX series with engines in the 115–204 hp bracket included front axle suspension, AutoSpeed transmission with push-button controls and hydraulic cab suspension.

229. A twenty-four forward and twelve reverse creep speed box was optional for the 53–102 hp McCormick CX with a twenty-four forward and reverse speed Synchro shuttle transmission.

230. The McCormick MTX tractor range had 118–176 hp turbocharged engines, four-speed powershift transmissions and luxury cabs with optional air conditioning.

Mercedes-Benz

Karl Benz and Gottlied Daimler, who were pioneers of the motor car, had engineering factories in different parts of Germany. Benz founded his business in 1871 and Daimler established his company in 1882 after working with Nicolaus Otto who developed the four-stroke cycle engine. Mercedes was the daughter of an important Daimler dealer in Germany and her name was used for a Mercedes-Benz car that took part in a 1901 motor race but the name was not used for Daimler-Benz vehicles until the two companies merged to form Mercedes-Benz in 1926.

The first Benz 40 and 80 hp tractors with a sprung front axle and a canopy over the driving seat appeared in 1919, followed a few years later by a 30 hp tractor with a twin-cylinder diesel engine. The 20 hp OE tractor introduced in 1928 was the first real Mercedes-Benz farm tractor. It had a single-cylinder water-cooled horizontal hot bulb diesel engine, three forward gears and one reverse, a gear drive transmission and an 8-gallon fuel tank.

With World War Two at an end a company called Gebruder Boehringer, a machine tool manufacturer at

Goppingen in Germany, introduced the Universal-Motor-Gerat with a 25 hp Mercedes diesel engine. The name was shortened to Unimog (page 262) when Daimler-Benz acquired Boehringer in 1950 and within ten years at least 50,000 Unimogs had been made at Gaggenau. Although Mercedes-Benz is well known for the Unimog they also introduced the four-wheel drive MB-trac in 1971.

The first four-wheel drive Mercedes-Benz MB-trac 65/70 with four equal-sized wheels launched at the 1971 DLG Show was equally suitable for fieldwork or road haulage. The 65/70 with a four-cylinder Mercedes engine rated at 65 or 70 hp, depending on model, was a systems tractor with a front and rear power take-off shaft and hydraulic linkage and a rear load-carrying platform. The MB-trac specification included a twelve forward and twelve reverse gearbox with a top speed of 25 mph, diff-locks on both axles, coil spring suspension on the front axle and a passenger seat in the cab. The more powerful MB-trac 95/105, introduced in 1974 was similar to the earlier model and the MB-trac 1100 and 1300 with 110 and 125 hp six-cylinder Mercedes engines appeared in

231. The Mercedes-Benz MB-trac 65/70 was launched at the 1972 German Trade Fair.

232. The bi-directional MB-trac 1500 (right) with coil spring front axle suspension was the most powerful Mercedes-Benz tractor in the early 1980s. A sales leaflet explained that the 80 hp MB-trac 1000 pulling the trailer had 'easy-to-use controls in a spacious cab with plenty of room for the driver and co-driver'.

1976. The 10,000th Mercedes-Benz MB-trac was made at Gaggenau in 1979.

The 150 hp MB-trac 1500 launched in 1980 was followed by the 85 hp MB-trac 900 in 1982 and the 95 hp MB-trac 1000 in 1983. By 1986 Mercedes-Benz (UK) Ltd at Milton Keynes were offering British farmers five models of the MB-trac from the 75 hp 800 to the 1500 with 150 hp under the bonnet. Early 1980s MB-tracs were designed to have an even weight distribution on the four equal-sized wheels. Standard features

233. MB-tracs including the six-cylinder 136 hp 1400 were marketed in the UK in the late 1980s.

included a twelve forward and reverse gearbox with a top speed of 25 mph, a diff-lock on both axles and hydrostatic steering.

Daimler-Benz and K-H-D formed Trac-technic in 1987 and Watveare Ltd became the UK distributor for the hand-built four-wheel drive MB-tracs. A new 110 hp model announced in 1987 increased the range to seven models from the 75 hp MB-trac 800 to the 156 hp MB-trac 1600. The engines were improved and a new fast response electronic hydraulic draft control system with greater sensitivity to soil conditions was a new option for

the 125 and 150 hp MB-trac 1300 and 1500. There were seven MB-tracs on the UK market in 1990 with 80–160 hp engines. The smaller models had a sixteen forward and eight reverse gearbox which included six creep speeds and the 1300, 1400 and 1600 had a sixteen-speed synchromesh gearbox with a reverse shuttle and eight creep speeds. They also had an electrically operated weight transfer system controlled from the cab which transferred the weight of a front- or rear-mounted implement onto the front or rear wheels to help reduce wheelslip and improve fuel economy.

The new flagship bi-directional MB-trac 1800 with a smart metallic light green livery and a turbocharged and intercooled six-cylinder 180 hp engine was launched at the 1990 Royal Show. A reverse console in the cab made it possible to use the 1800 with a rear-mounted forage harvester in the same way as a conventional self-propelled machine. The last MB-tracs were made in 1991 although Reed UK at Melksham in Wiltshire imported a range of similar 80, 50 and 180 hp tractors made by Werner GmbH. The new 200 hp Trac 200 with a six-cylinder Mercedes engine and forty forward and reverse powershift transmission was added to the existing model range in 2001.

234. The seat and steering wheel were offset on the 'Vision-Lined' Minneapolis-Moline Model Z first made in 1937.

Minneapolis-Moline

The Moline Plow Company, the Minneapolis Threshing Machine Company and the Minneapolis Steel & Machinery Company merged in 1929 to form the Minneapolis-Moline Power Implement Company. The Moline Plow Company introduced the 18 hp Model D motor plow in 1915, and the Twin City 6–30 tractor was being made by the Minneapolis Steel & Machinery Company in 1920. The Minneapolis Threshing Machine Company was building the 44 hp MTM 24–44 tractor with chain-operated steering and a distinct likeness to a traction engine in the early 1920s. Following the amalgamation Minneapolis-Moline made Twin City tractors until 1938.

The two/three-plough Minneapolis-Moline Model Z tractor with a 31 hp four-cylinder engine and five-speed transmission launched in 1937 was 'Vision-Lined' with a tapering bonnet. The fuel tank and steering wheel were offset to the left to give the driver a clear view when working in rowcrops. Like other American tractors of the day there were several variants of the Model Z with its Prairie Gold paintwork, red wheels and red radiator grille. The ZTS was the standard version, the rowcrop ZTU had vee-front wheels, the ZTN had a single front wheel and the XZTI was an industrial tractor. Unlike

modern glossy brochures some Minneapolis-Moline sales literature gave very detailed information and the model Z leaflet even illustrated many engine and transmission parts including a valve spring, rocker gear and the camshaft.

Sale Tilney & Co at London, Essendine and at Wokingham in Berkshire imported their first M-M tractors in time for the 1938 Smithfield Cattle Show and exhibited the Minneapolis-Moline UDLX at the 1939 Royal Show. Early M-M tractors sold in the UK included the standard ZTS model, the GT, the RT with disc brakes and the UDLX with a comfort cab complete with a radio and clock. Advertised as 'the mighty master of all jobs' the 48 drawbar hp GT, the largest M-M tractor at the time, had a four-cylinder water-cooled engine. The cylinder block was cast in two parts and there was a choice of high- or low-compression heads for the petrol or paraffin engines. The GT with a four forward and one reverse gearbox, a top speed of 9½ mph and a transmission band brake was made from 1938 until the 55 drawbar hp GTA replaced it in 1942. The GTA was similar to the GT with internal expanding brakes and the option of pneumatic tyres or steel wheels but an enclosed flywheel and a channel-section engine chassis

235. The late 1930s Minneapolis-Moline Model R had a 'vision-lined quick-removable comfort cab' with a deep bucket-type seat.

replaced the earlier heavy cast-iron framework. A Sale Tilney advertisement for the M-M Model U in 1943 explained that the tractor produced enough traction with its four-cylinder engine to pull a four- or five-furrow plough. The five-speed transmission provided speeds from 2½ mph and the tractor with its light and easy steering turned in a 12 ft radius.

Sale Tilney and Minneapolis-Moline formed M-M (England) Ltd in 1946 to manufacture Minneapolis-Moline combine harvesters, saw benches, winches, hammer mills and other products in the UK. The UDS tractor was also assembled at Wokingham, some of the parts being made locally with the remainder brought in from America. The UDS had a 46 hp Dorman engine, the British-built UDM had a 66 hp Meadows engine and both tractors had a 24-volt electric starter. Otherwise the tractors had the same five forward and one reverse transmission, expanding shoe brakes, power take-off and a belt pulley.

M-M (England) Ltd went into receivership in 1949 but Sale Tilney was still importing the American-built tractors until the mid-1940s when the distinctive yellow tractors with red wheels disappeared from the British market. Sale Tilney also imported New Holland balers and this arrangement continued until 1955 when the franchise was taken over by Western Machinery, based in Devon.

The Cleveland Tractor Co, which made Cletrac crawlers and the General GG wheeled tractor, was acquired by Oliver Farm Equipment in 1944, who in turn were taken over by Minneapolis-Moline in 1951. Oliver retained the Cletrac crawler range while BF Avery & Sons bought the production rights for the wheeled General GG tractor. The four-cylinder Avery Model R made at the time became the M-M Model BF. Minneapolis-Moline tractors were no longer available in the UK but they were still being made in America and by 1953 there were six basic models ranging from the 60 hp four- or five-plough Model G introduced in 1947 to the one-plough Model V previously made by BF Avery. An advertisement in 1953 explained that the price of a tractor in 1940 was 'equivalent to the market price of nine beef cattle but in 1953 farmers can get a much better (Minneapolis-Moline) tractor for less than the market price of four beef cattle'. A similar comparison today is difficult to imagine. A testimonial in the same advertisement stated that 'I'm farming with an M-M because I've found out just how much money it can make for me.' These words were obviously uttered by an American farmer, as most of his British counterparts would never admit to making a profit!

M-M made the high-power Super 90, 95, Super 95 and 97 for Massey Ferguson from 1958 to the early 1960s. The White Motor Co, which already owned Oliver and Cockshutt, bought M-M in 1963. The White Motor Co continued tractor production under the Oliver, Cockshutt and M-M names until 1974 when they were all sold as White tractors.

236. The Minneapolis-Moline Model U had a four-cylinder paraffin engine.

237. Moffett Multi-Function Tractors were made in Ireland.

M-M made the high-power Super 90, 95, Super 95 and 97 for Massey Ferguson.

Moffett

Moffett Engineering, already established as a forklift truck manufacturer at Dundalk in Ireland, introduced the first Moffett Multi-Function Tractor (MFT) in 1991. The bi-directional MFT, based on a Massey Ferguson 390T tractor skid unit with a 90 hp Perkins engine, was a four-wheel drive tractor with a rear-mounted 2 ton capacity industrial loader. The specification included a Massey Ferguson twelve-speed gearbox with a reverse shuttle, an optional four-speed torque converter and wet disc brakes on the rear axle. Three-point linkage and two-speed power take-off were standard and an optional front linkage was available. The driving seat and controls, which could be reversed in a matter of seconds, gave equal visibility in both directions.

The 120 hp MFT based on a Massey Ferguson skid unit and with an improved cab replaced the original model in 1994. It had a twelve-speed reverse shuttle gearbox and was designed for ploughing, forage harvesting and other heavy land work. The Moffett MFT7840 with a 100 hp six-cylinder New Holland 7840SL skid unit superseded the MF-based tractor in

1995. The 7840 with a twenty-four-speed SynchroShift gearbox with Dual Power and a clutchless reverse shuttle had a de-mountable parallel linkage rear loader. Standard features included top link sensing hydraulic linkage and two-speed power take-off. Like earlier Moffett Multi-Function tractors the 7840 could be used either as a conventional tractor or, after spending a few minutes reversing the driving controls and attaching a rear-mounted loader, for material handling duties.

Morooka

Yuasa Warwick Machinery introduced a range of 80–325 hp Morooka rubber-tracked crawlers to British farmers at the 1993 Royal Show. Established in Japan in 1958 Morooka, a manufacturer of earth-moving machinery, fork trucks and loaders, was one of the first companies to make rubber-tracked crawlers. The MK-100 with a 325 hp Cummins engine was the largest of the nine-model range of Morooka crawlers available in 1995. The smallest MK-100 had a 100 hp Japanese Hinto engine, the others had a Komatsu, Mitsubishi or Perkins power unit.

Robert H Crawford & Son at Boston in Lincolnshire, who were appointed Morooka distributors in 1997,

238. The Japanese Morooka rubber-track crawlers were introduced to British farmers in 1997.

included eight models in the 65–325 hp bracket with hydrostatically driven reinforced rubber tracks steered with two levers in their price list. Komatsu diesel engines provided the power for the smaller Morooka tractors while others had a Cummins power unit. Standard features included category II hydraulic linkage, twin-speed range hydrostatic transmission and dual-speed power take-off. Optional equipment included category III linkage on the 250 and 325 hp models and a 1,000 rpm power take-off for the largest model in the range. The Morooka RHC180 and RHC250 rubber track crawlers, current in 2001, had a hydrostatic transmission and the model number indicated the horsepower output of the six-cylinder engines.

Muir-Hill

Mr Muir and Mr Hill were making shunting engines based on Fordson tractors for railway trucks at Manchester in the early 1920s and by the end of the decade Muir-Hill were also using Fordson power units for a range of dumper trucks. Winget, well known for their concrete mixers, bought Muir-Hill in 1959 and production of dumper trucks and loader shovels continued at Manchester until the early 1960s.

Muir-Hill had moved to Gloucester when the four-wheel drive M-H101 designed for agricultural and industrial use was launched in 1966. The tractor had a 108 hp six-cylinder Ford industrial engine and four equal-sized wheels and a low-speed model for drainage and forestry work was added in 1967. The M-H101 cost under £3,000 and like many of its four-wheel drive competitors it had a Ford 5000 gearbox with eight forward gears and a top speed of 20 mph. Muir-Hill also used Ford transmission and hydraulic systems but they made their own transfer box and drive shaft to the front axle. The front and rear wheel track was adjustable from 64 in to 80 in and front-wheel drive could be disengaged when necessary. Other features included oil-immersed disc brakes on both axles, power-assisted steering and an independent hand lever-operated clutch for the power take-off.

Winget and Muir-Hill joined the Babcock and Wilcox group in 1968 and the M-H110 with a six-cylinder 110 hp Perkins engine and the M-H161 with a 163 hp V8 Perkins engine were launched in 1969. With a ten-speed gearbox, hydrostatic steering, twin exhausts and detachable cab the M-H161 was the most powerful farm tractor on the British market in the late 1960s. An improved M-H101 with a 120 hp Ford engine and full hydrostatic steering appeared in 1971 but within twelve months Muir-Hill launched the Series II tractors with 110, 120 and 163 hp engines and improved flat floor cabs. The M-H111 with a 110 hp Perkins engine

239. A six-cylinder 108 hp Ford diesel engine provided the power for the Muir-Hill 101.

replaced the M-H110 and the Muir-Hill 121 with the same 120 hp Ford engine and gearbox superseded the 101. The Series II 161 with twin exhaust pipes remained in production until 1975 when it was replaced by the M-H171 with a 170 hp V8 Perkins engine, lower link sensing hydraulics and new cab.

The Series III M-H111, 121, 141 and 171 tractors announced in 1978 were similar to previous models except for a new sound-insulated Spacecab with a tinted glass windscreen. The new cab with an

240. Sales literature explained that the 'giant Muir-Hill 161 with a 163 hp Perkins engine had the same unbraked turning circle as a family car'.

opening rear window and a push-button radio had a heating and ventilation system that changed the air five times every minute. The M-H141 with a six-cylinder turbocharged 143 hp Ford or a 132 hp Perkins engine was the new addition to the Series III range. The eight forward and two reverse Ford gearbox with optional Dual-Power was used for the 111, 121 and 141 but the eight-cylinder 170 hp M-H171 had a ten forward and two reverse gearbox, hydrostatic steering and self-adjusting oil-cooled disc brakes on both axles.

Muir-Hill tractor production came to an end in 1982 when Babcock sold off its construction equipment division. Following a period under various different owners, including Sanderson Forklifts and Aveling-Barford, the business was acquired by Lloyd Loaders at Mytholm in Yorkshire. The new owners introduced the Myth-Holm 131 in 1989 and this tractor remained in small-scale production for about six years. Very similar to its Muir-Hill ancestors it had a six-cylinder 130 hp Ford engine, a sixteen forward speed Dual Power transmission and an air-conditioned cab.

New Holland

Abram Zimmerman started a farm implement repair business at New Holland, Pennsylvania in 1895 and by the end of the century he was making a range of barn equipment. Stationary engines were added in the early

1900s and the New Holland Machine Co was making automatic twine-tying pick-up balers in the late 1930s. The Sperry Corporation purchased New Holland in 1947 and by the mid-1950s the company was making a wide range of farm machinery including balers, bale loaders, forage harvesters and manure spreaders. New Holland purchased a major interest in Claeys factory in Belgium in 1964 and combine harvesters were then added to the product range.

The Ford Motor Co acquired Sperry New Holland in 1986 and Ford New Holland bought the Versatile Manufacturing Ltd of Canada in 1987. The next change came in 1991 when Fiat bought Ford New Holland and the headquarters, renamed New Holland Geotech, were established at Brentford in Middlesex. New Holland Geotech became New Holland in 1993.

Fiat tractors were still made in Italy and Ford tractor production continued at Highland Park in America, Antwerp and Basildon until the first mechanically identical blue New Holland Ford 70 series and terra cotta Fiatagri G series tractors were launched in 1994. The Fiatagri G170, G190, G210 and G240 model numbers indicated the horsepower of the turbocharged engines and the same power units were used for the Ford 8670, 8770, 8870 and 8970 tractors. An eighteen forward and nine reverse electro-hydraulic Powershift transmission with microprocessor controls was standard

241. The driving console on 135 hp bi-directional New Holland TV140 could be reversed in a few seconds.

on the four-wheel drive tractors. The Super-Steer front axle with a 65-degree steering angle compared with the usual 50-degree angle was an optional extra. The Fiat G series and Ford 70 series were the last new models to have the Fiatagri or Ford oval badge on the radiator grille and within a year the New Holland blue leaf logo was used for the terra cotta and the blue tractors. New Holland launched the 65–90 hp Ford 35 series and Fiat L series together with the 100–160 hp Ford 65 series and Fiat M series tractors in 1996. Farmers needing even more power could buy a 360 hp Versatile 9682 in New Holland blue livery. The 9682 was one of four high horsepower tractors made at the time at the Versatile factory in Canada.

New Holland announced the medium-powered TS90, 100 and 110 and three TNF fruit tractors in 1997. The 80, 90 and 100 hp TS tractors with a twenty-four speed forward and reverse transmission replaced some of the earlier 40 series models but the popular 7840 and 8340 remained in production. The New Holland Super-Steer front axle with an improved 76-degree steering angle was a feature of the new 65 hp, 76 hp and 88 hp TNF fruit tractors. The TS range was extended with the launch of the six-cylinder 100 hp TS115 at the 1998 Royal Smithfield Show. The new model with a twenty-four speed reverse shuttle transmission replaced the 100

hp 7840. The articulated bi-directional New Holland TV140 tractor with a six-cylinder 135 hp engine was introduced in 1998 to assess potential sales in the UK. The driving console in the TV140 cab could be rotated in a matter of seconds in readiness to drive the tractor in either direction.

There were ten different models of New Holland tractor in 2000. The smallest four-wheel drive TNS and the two- and four-wheel drive TND had 50, 60 and 92 hp engines and a sixteen forward and reverse transmission. The four-wheel drive 8670 to 8970 had 170–240 hp power units and the flagship Canadian-built New Holland Versatile 9682 with 360 hp under the bonnet had a twelve forward and two reverse speed gearbox.

Three new models of the 190–240 hp 70A series with improved Super-Steer and a fully locking front differential were added in 2001. The 165 hp TM165 with Power Command transmission was added to the 100–160 hp TM range originally launched in 1999. A 'Classic' version of the 115 hp TM125 with a mechanical gearbox, launched in 2001, was described as an ideal tractor for the occasional tractor driver. The 177 hp TM175 and 194 hp TM190 announced in 2002 had an engine speed management system used to select a constant engine speed that would be maintained regardless of the load on the tractor.

The New Holland 231, 258 and 283 hp TG series launched in 2003 had six-cylinder turbocharged twenty-four valve Tier 2 engines, an eighteen forward and four reverse powershift transmission and a spacious cab with automatic climate control. Five new TVT tractors with six-cylinder 137–192 hp engines and CVT variable transmission appeared in 2004 along with four new TL-A and TS-A series tractors. An optional factory-fitted front-end loader was available for the TL-A tractors with 72–100 hp four-cylinder engines and low-profile cabs. The TS-A models with 101–136 hp common rail fuel injection engines had an Electro Command semi-powershift

242. The three New Holland TG series tractors launched in 2003 had 231–283 hp twenty-four valve six-cylinder Tier 2 engines.

243. Launched in 2004 the New Holland TVT 190 had a constantly variable transmission with no fixed gear ratios and a top speed of 30 mph.

transmission. There were fourteen different ranges of tractor in the 2005 New Holland price list. They included ground care, orchard and vineyard models, medium-power tractors for livestock farms and 100 hp-plus models for the larger arable farm. A 94 hp TK-A crawler tractor with steel tracks, a 94 hp turbocharged Iveco engine and an eight-speed forward and reverse gearbox was also available on the UK market.

Newman

Newman Industries, manufacturers of dynamos and electric motors at Yate near Bristol, acquired Grantham Productions in Lincolnshire where the three-wheeled Kendall tractor was made in 1946 and 1947. Following a major refurbishment of the Grantham factory Newman Industries introduced the tricycle-wheeled Newman AN3 and AN4 tractors in 1948. Horizontally opposed air-cooled Coventry Victor petrol engines were used for the 10 hp Newman AN3 and for the 12 hp Newman AN4 which was made for farms where more power was required to deal with difficult soil conditions. Otherwise the Newman and earlier Kendall tractors were very similar with a single-plate clutch, a three forward and one reverse gearbox with a top speed of 10 mph, independent brakes and a swinging drawbar. Optional equipment included a mid-mounted toolbar with manual or hydraulic lift, a rear toolbar, power take-off and a belt pulley.

The improved three-wheel Newman WD2 with a Coventry Victor water-cooled single-cylinder diesel engine appeared in 1949. Flow and return water pipes connected the mid-mounted engine to the radiator and cooling fan at the front of the tractor. The WD2 had a single-plate dry

clutch, a conventional transmission with a three forward and one reverse gearbox, a top speed of just under 9 mph and 1 mph in reverse and independent brakes. The rear wheels, which could be moved in or out on their axles, provided an infinitely variable track setting anywhere between 42 and 72 in. The standard tractor cost £330 but with the optional power take-off, belt pulley and hydraulic linkage the price was £410.12s which was considerably more expensive than the £335 price tag for a Ferguson TED20.

The four-wheeled Newman Model E2 with a 12 hp twin-cylinder Petter water-cooled diesel engine and a four forward and one reverse gearbox appeared in 1951. The engine was started by hand with the aid of a decompressor and a fuel priming lever. The basic model with adjustable wheel track settings from 42–54 in at the front and up to 72 in at the rear cost £430. Optional extras included a rear-mounted belt pulley for £10, a 1½ in diameter power take-off shaft for £18 and a hydraulic

244. The Newman with its high ground clearance and mid-mounted toolbar was an ideal rowcrop tractor.

system with a gear pump in the transmission housing added an extra £46.10s to give a total price of £504.10s. A swinging drawbar was standard and the tractor could be used with front, mid-mounted or rear toolbars raised and lowered with a hand lever or the optional hydraulic system. Its high ground clearance and mid-mounted toolbar made the Newman E2 an ideal tractor for rowcrop work.

Northrop

The Chaseside company based in Middlesex, which manufactured loading shovels and railway shunters based on Fordson tractors, was taken over by British Northrop at Blackburn in the late 1950s. British Northrop made machinery

245. The Northrop 5006, launched in 1967, had a 90 hp Ford diesel engine and an eight forward and two reverse speed gearbox.

for the weaving industry and it was the Chaseside expertise that led to the introduction of the equal-sized four-wheel drive Northrop 5004 tractor in 1965. Built mainly with Ford 5000 components the Northrop had a 67 hp four-cylinder engine, an eight-speed gearbox or optional Select-O-Speed box and a central drive shaft from a transfer box to the front axle.

The head office of Northrop Tractors was at Ware when a turbocharger was added to the 5004 engine which increased the rated power to 85 hp. The Northrop 5006 launched in 1967, also based on the Ford 5000 tractor, had a six-cylinder 90 hp Ford diesel engine and an eight forward and two reverse gearbox with a top speed of 20 mph. The specification included epicyclic reduction gears, diff-locks, hydraulically operated disc brakes on both axles and the track width was adjustable from 56 in to 80 in. The 5006 complete with category II hydraulic linkage and power take-off cost £2,900 but with only a handful made the tractor was discontinued in 1967 and within twelve months British Northrop sold the Chaseside part of business to JCB.

Nuffield

The Nuffield tractor and the Morris motor car were both named after William Morris, later Lord Nuffield. The Morris Motor Co decided to enter the farm tractors market when the demand for military vehicles diminished at the end of World War Two. The Agricultural Division of Wolseley Motors chose the 1948 Royal Show to launch the Nuffield M4 and tricycle-wheeled M3 tractors with 38 hp vaporising oil engines based on a four-cylinder Morris Commercial side-valve power unit. The relatively advanced design of the Nuffield tractor included a five forward speed and one reverse gearbox and electric starting and the rear wheel track width was adjusted by sliding the wheels in or out on their axles. Prices started at £495 for the M4 and £487.10s for the tricycle-wheeled M3 but hydraulic linkage with two auxiliary service points and power take-off added another £80 to the bill. The optional single front wheel, mainly intended for the North American market, was an unusual feature for a British-built tractor. The single front wheel conversion kit for the M4 cost

246. The Nuffield Universal was launched at the 1948 Royal Smithfield Show.

£37 and the kit needed to convert the M3 to M4 was £46.10s.

A sales leaflet described the Nuffield as 'the newest aid to power farming'. It was explained that, 'Much thought has been given to driver comfort shown by the provision of a deep-cushioned, rubber-sprung seat, good vision of work, light steering, conveniently grouped controls and a low floor with an easy driving or standing position'. The Nuffield PM4 and tricycle-wheeled PM3 introduced in 1950 had 38 hp petrol engines.

The first diesel-engined Nuffield Universal appeared in 1950. The DM4 with a 48 hp Perkins P4 (TA) engine cost just under £670 compared with £550 for the M4 with a vaporising oil engine. Morris Motors became part of the British Motor Corporation in 1952 but the Perkins engines were used until a new 45 hp BMC diesel engine replaced it in 1954. The renumbered Nuffield 4DM was advertised as a tractor with 'even greater fuel economy, smoother running, easier starting and better cooling at a very reasonable price'. An optional independent power take-off and hydraulic system with a dual clutch was introduced in 1956. The transmission clutch was

controlled with a pedal and a hand clutch was used for the power take-off. The Nuffield 4DM became the Universal Four when the 37 hp Universal Three with a three-cylinder BMC direct injection diesel engine and a five forward and one reverse speed gearbox appeared in 1957.

Both tractors had conventional dished rear wheel centres for rear track width adjustment while the earlier sliding hub arrangement on the rear axle was optional on the Universal Four. A rather sparse specification required a long list of optional extras to make the Universal Three a serious competitor for the Massey Ferguson 35 or Fordson Dexta. However, electric starting, steering brakes and a parking brake, a swinging drawbar, oil for the engine and transmission and a licence holder were included in the basic price of £555. Optional category I and II hydraulic linkage and independent power take-off cost £122.10s but the diff-lock only added £5 to the bill. Electric lighting, horn and number plates cost ten guineas and the extra comfort provided by a de luxe seat with a cushion added another £3.2s.6d. For a relatively low basic price of £610 the Nuffield Universal Four had a BMC diesel engine and a

247. *The Nuffield Universal Four superseded the Universal 4DM in 1957.*

five forward and one reverse gearbox but an independent power take-off and hydraulic system were some of the extras needed for the Universal Four to match the competition.

Technical information published in the late 1950s concerning tractor horsepower was often confusing and Nuffield Universal Four sales literature which gave six different horsepower ratings was typical. The list started with the maximum 56 SAE bhp (bare engine brake hp measured at the flywheel), with the engine driving the water pump and cooling fan engine power was reduced to 53 Din bhp at 2,000 rpm and 39.6 bhp when the engine ran at 1,400 rpm. Din hp is a European standard for net horsepower. Drawbar power was given as 34.8 hp and 45.8 at 1,400 and 2,000 engine rpm respectively and the belt pulley developed 52.1 hp when running at the British Standard belt speed of 3,100 ft/min.

The Universal name was dropped when the Nuffield 342 and 460 superseded the Universal Three and Four in 1961. The first figure related to the number of cylinders and the other two indicated the approximate engine horsepower. Even in 1961 the price of £595 for the 342 still excluded hydraulic linkage and power take-off. The basic Nuffield 460 with hydraulics and a power take-off cost £650 and the de luxe model priced at £810 had an independent power take-off and depth-control hydraulics. Tractor production was transferred from Birmingham to a new BMC truck factory at Bathgate in Scotland in 1961. The 10 series of tractors announced in 1964 were the first Nuffields to have the power take-off and hydraulic linkage with interchangeable dual category lower link ball ends as standard equipment. The de luxe models had an independent power take-off shaft. The 57 hp 10/60 with ten forward speeds was faithful to its model number but the 10/42 also with ten forward gears was rated at 42 hp. Disc brakes were a new feature which BMC sales literature explained were just as efficient as expanding shoe brakes. Power-assisted steering and a rear-mounted pto-driven belt pulley were optional extras for both tractors. The original Nuffield Universal sliding rear wheel hubs which gave stepless wheel track settings were optional for the 10/60.

The British Motor Corporation forecast a 'new era in farming economy' when they launched a 15 hp Mini-Tractor at the 1965 Royal Smithfield Show. An

248. The Nuffield 10/42 was claimed to be the first British-built tractor with ten forward gears.

249. A 1965 advertisement in a farming magazine claimed that 'the Nuffield BMC Mini-Tractor would start a new era of Mini-mechanisation'.

250. The Nuffield 4/65, which replaced the 460 in 1967, had a re-styled bonnet, wide topped wings and a 15 gallon fuel tank located in front of the radiator.

advertisement explained that, 'The light, compact, ultra-manoeuvrable and ultra-economical BMC Mini-Tractor is capable of all but the heaviest tasks and it will start a new era of Mini-mechanisation.' The 'entirely new and ingenious prime mover' with its own range of implements would 'cut farm economics down to size and stem the growing trend towards bigger and bigger tractors'. In reality the BMC Mini, made by Nuffield but without their name badge, was similar to the Ferguson TE20 introduced some twenty years earlier. A four-cylinder indirect injection engine with a CAV rotary-type fuel injection pump and a heater coil in each cylinder for cold starting provided the power for a nine forward and three reverse gearbox with a top speed of just under 13 mph. The basic tractor with disc brakes, diff-lock, parking latch, electric starter and tractor meter cost £512.10s. The HPU model with hydraulic linkage, a front-loader kit and lights was another £62.10s. The BMC Mini-Tractor was not a great success and was relaunched in 1968 as the Nuffield 4/25 with a more powerful four-cylinder, indirect injection 25 hp diesel or optional 25 hp petrol engine.

Improved and restyled 3/45 and 4/65 tractors replaced the Nuffield 10/42 and 10/60 in 1967. The same power units were used but the 4/65's higher engine speed increased the power output to 65 hp and the improved hydraulic system had a double-acting top link. The tractor had two hydraulic pumps, one supplied oil to the lift cylinder and external rams and a smaller pump was used for the draft control system. The 10/42 and 10/60 gearboxes with top speeds approaching 20 mph were retained but power steering, still considered a luxury item, was an optional extra. The drag link was put inside the main frame to tidy the steering linkage and keep it out of the way of side-mounted implements. Prices ranged from £765 for the standard 3/45 with hydraulics to £1,044.10s for the de luxe 4/65 with independent power take-off. Four-wheel drive conversions were popular by the late 1960s and Bray Construction Equipment followed the trend with four-wheel drive versions of the 4/65 and the earlier 10/60.

The BMC Nuffield tractor operation became part of British Leyland in the late 1960s. The Nuffield name was retained until December 1969 when the poppy red 4/25, 3/45 and 4/65 became the Leyland 154, 344 and 384 with a new two-tone blue colour scheme.

251. The two-wheel drive 230 hp OnTop fast tractor had a top speed of 45 mph.

OnTop

Scotlon Flotation Equipment Ltd exhibited OnTop lime spreaders and slurry tankers at the Royal Show in 1984. The 120 hp OnTop lime spreader with twenty forward gears carried up to 8 tons of lime and the 152 hp slurry tanker with a creep speed gearbox had an application rate of 1,000–2,000 gallons an acre.

OnTop Tractors at Balgowan near Perth in Scotland made their first two- and four-wheel drive fast tractors in the late 1980s. A combination of new and recycled ex-military running rear was used to manufacture the tractors designed to pull up to 24 tons at speeds of 30–45 mph. The two-wheel drive 800 series OnTop Cruiser and four-wheel drive Trekker were exhibited at the 1991 Royal Show. The Cruiser with a 230 hp six-cylinder turbocharged Leyland diesel engine, a ten forward and eight reverse constant mesh gearbox and a top speed of up to 40 mph had an auxiliary hydraulic system, split circuit air brakes and optional hydraulic linkage. The specification of the two- and four-wheel drive OnTop Trekker included a 205 hp

turbocharged Bedford diesel engine, a twin-range synchromesh gearbox with twelve forward gears and a top speed of 45 mph, power-assisted steering and air-operated brakes combined with a trailer braking system. Sales literature explained that the Trekker was built in a standard format and could be 'dressed' to suit each customer's requirements. Optional equipment for the Trekker and Cruiser tractors included a dual speed power take-off, live hydraulic system, twin rear wheels, a pick-up hitch, air conditioning, a hand throttle and a ground speed power take-off for use with power-driven trailer axles.

OTA

The name of the OTA tricycle-wheeled tractor, cast in large letters on the radiator grille, was derived from the initials of Oak Tree Appliances at Coventry. After an appearance at a local agricultural show the OTA was launched at the 1949 Royal Smithfield Show. The first OTA tractors were painted red and yellow but this was soon changed to blue. The OTA with a petrol engine and hydraulic linkage cost £246.10s. A Beccles vaporising oil conversion kit cost £10.10s, a four-speed power take-off shaft with a 10 in belt pulley was £22.10s and a pair of wheel strakes was an extra £11.6s.8d.

The OTA had a 10 hp water-cooled, side-valve Ford industrial engine with coil ignition and an electric starter. There was a starting handle dog on the crankshaft pulley and a second dog at the back of the tractor was attached to the end of the shaft from the gearbox to the final drive gears. This enabled the driver to hand crank the engine from the rear which was just as well because the engine radiator

252. Oak Tree Appliances at Coventry made the OTA tractor between 1949 and 1953.

obstructed the crankshaft pulley starter dog so it could only be started from the back of the tractor. The main gearbox had to be in gear when starting the tractor but it was very important to make sure that the high/low lever was in neutral before using the starting handle. A three forward and one reverse speed Ford gearbox combined with a high/low ratio lever provided six gears with a top speed of 15 mph and two in reverse. The OTA had a worm and wheel final drive and oil for the live hydraulic system was provided by a twin-cylinder piston pump belt driven from the engine crankshaft pulley. The steel channel chassis was set at an angle to raise the front of the tractor to accommodate the single front wheel turned by a cable from the steering wheel. The resultant high ground clearance gave an excellent view when using an underslung toolbar in rowcrops.

A redesigned bonnet and a sheet metal radiator without the large cast-iron OTA badge were the most obvious changes on the Mk II tractor introduced a

year or so later. Some Mk II tractors, also known as the 5000 series, had an improved transmission system and power take-off. The 1951 Royal Smithfield Show was chosen to launch the four-wheeled OTA Monarch and except for a new front axle with a conventional steering linkage and hinged bonnet it was very similar to the tricycle-wheeled model. Wheel track settings could be varied between 42 and 60 inches and various implements were available for the mid- and rear-mounted tool frames.

Singer Motors bought the manufacturing rights for Oak Tree Appliances tractors in 1953. The three-wheeled OTA was discontinued but the Monarch was made until 1956 when the Rootes Group acquired Singer Motors.

Chapter 7

Paramount – Russell

Paramount

Ernest Doe coupled two tractors together to make the 100 hp-plus Doe Triple D and it was possible to split them back into two separate tractors but this was rarely if ever done. However, it was relatively easy to reverse the procedure when the Paramount Dual Tractor Kit was used to couple two tractors together. The Paramount Dual Tractor Kit designed by a Berkshire farmer and made by Paramount Engineering at Coventry in 1968 could be used to link any two popular models of tractor together with the minimum of expense. Sales literature explained that when the power of a four-wheel drive tractor was required the Paramount kit was 'as simple to use as attaching a trailer and the driver can operate all the controls including throttle, clutch, gear lever, diff-lock and hydraulics while seated on the front

tractor'. The front axle and steering linkage were removed from the rear tractor and replaced with a ring hitch. The pick-up hitch on the leading tractor was used to couple the two tractors together and the Paramount Dual Tractor unit was ready for work.

A sales leaflet explained that the extremely low price and its ability to increase the work output of conventional tractors made the Paramount Dual Tractor Kit a must for every farm. Better still, there was no need to invest capital in heavy equipment which could only be used economically and effectively for a very short period of the year. However, farmers were not impressed by the Paramount publicity and the conversion kit was not a commercial success.

253. Two identical tractors were not required when the Paramount Dual Tractor Kit was used.

254. Porsche made some models of German Allgaier tractors, including the twin-cylinder AP 22 with an air-cooled engine, in the late 1950s and early 1960s.

Porsche

Dr Ferdinand Porsche, designer of the Porsche sports car and the Volkswagen Beetle car also made the rear-engined Porsche Volkschlepper (people's tractor) with a twin-cylinder power unit in 1938 and an improved Volkschlepper with a twin-cylinder 40 hp Allgaier diesel engine appeared in the mid-1940s.

Allgaier, with a factory at Württemberg in Germany, bought the Porsche tractor division in 1946 and the first Allgaier R18 and R22 tractors with single-cylinder 18 and 22 hp hopper-cooled diesel engines appeared in 1947. They were very basic machines with an old-fashioned appearance compared with the Ferguson TE20 and other tractors of the day. Ferdinand Porsche renewed his interest with Allgaier in 1949 and this resulted in a change from hopper-cooled to air-cooled engines in 1952. The Allgaier AP17 with an air-cooled twin-cylinder engine appeared in 1952 and the 12 hp A111, launched in 1954, had a single-cylinder air-cooled engine, a four forward and one reverse gearbox and power take-off. There were short and long wheelbase versions of the A111 tractor which with its high

ground clearance and narrow waist provided maximum visibility for rowcrop work.

The Allgaier name was dropped in favour of Porsche in 1956 when the A122, A133 and A144 Porsche tractors were made in Germany. The smallest A122 with a 22 hp twin-cylinder air-cooled diesel engine had a five forward and one reverse gearbox. The 33 hp A133 and 44 hp A144 with three-cylinder and four-cylinder engines respectively had a fluid flywheel between the crankshaft and conventional single-plate clutch similar to that used on Fendt tractors. A 540 rpm power take-off was standard on all three models and the A144 also had a front power take-off shaft which ran at half engine speed.

The A111 became the P111 in 1956 and with the engine speed increased to 2,250 rpm in 1957 and rated at 16 hp, it was renamed the Porsche Junior L and the Porsche P122, P133 and P144 became the Porsche Standard, Super and Master in 1958. Porsche tractors were not available in the UK until 1959 when Eurotrac (Imports) Ltd at Dover were appointed sole

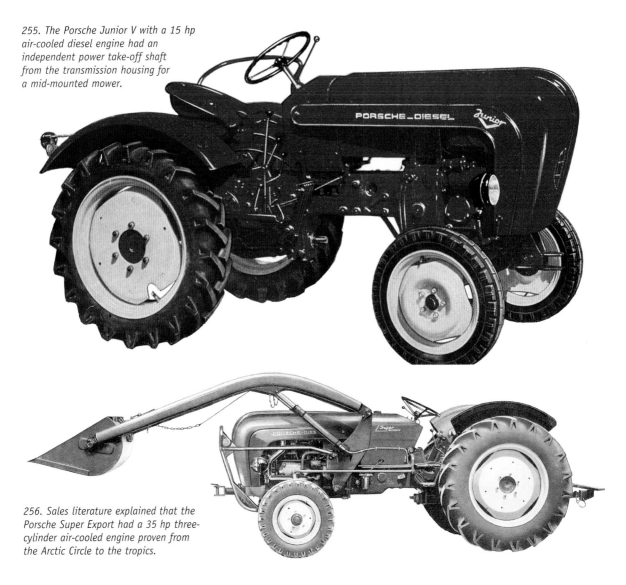

255. The Porsche Junior V with a 15 hp air-cooled diesel engine had an independent power take-off shaft from the transmission housing for a mid-mounted mower.

256. Sales literature explained that the Porsche Super Export had a 35 hp three-cylinder air-cooled engine proven from the Arctic Circle to the tropics.

distributors. Three models were available in the UK in 1962 including the Porsche Super Export, Standard and Junior V. Sales leaflets explained that tractors for the home market had large wrap-round mudguards but shell-type mudguards were used for the export market. The 33 hp Porsche Super Export with a £985 price tag had eight forward and two reverse gears, front-, mid- and rear-power take-off shafts and hydraulic linkage. The 22 hp Standard had live mid- and rear-power take-off shafts and hydraulic linkage and the 15 hp Porsche Junior V, with a similar specification, completed the range imported by

Eurotrac. However, their involvement with the German tractor maker was shortlived as Porsche tractors went out of production in 1964.

Power Take-Off

The power take-off shaft appeared in the late 1920s and many tractors had one by the mid-1930s. There were no real construction standards in the early days either for shaft dimensions or operating speed. British Standard Specifications for agricultural tractors introduced in 1948 specified a six-spline 1⅜ in diameter shaft running at 536 rpm plus or minus 10. Most manufacturers met these standards but the Ferguson TE20 series had a 1⅛ in diameter shaft. This caused a few problems with the increased range of power-driven implements on the market in the early 1950s but Lawrence Edwards & Co and a few other companies made an adaptor to convert the TE20 power shaft to the standard size.

Ground speed power take-off with the splined shaft running at a speed proportional to the forward speed was introduced in the mid-1950s and the Ferguson FE35 was one of the first tractors with this facility. The British Standard power take-off speed was changed to 540 rpm plus or minus 10 in 1958 and the American Society of Automotive Engineers (S.A.E.) introduced a second standard speed of 1,000 rpm for a twenty-one spline 1¾ in diameter shaft in the same year. The new 1,000 rpm standard was ignored for a while in the UK but it eventually became a standard feature on the more powerful tractors sold in Britain. The 1,000 rpm power take-off speed used to drive machines with a high power requirement soon became standard on most medium and high horsepower tractors. The dual speed power take-off had a high/low ratio gearbox with a 540 rpm six-spline shaft and a twenty-one splined shaft running at 1,000 rpm. The next development was the use of a six-spline power shaft in the 1,000 rpm setting with the engine running at rated speed and at a lower engine speed to provide an economy 540 rpm speed for pto-driven implements with a low power requirement.

Power take-off shafts were a major cause of farm accidents until farm safety legislation introduced in 1956 required the pto shaft to be guarded from the tractor to the first fixed bearing on the machine. The guard on the tractor had to be strong enough to support 250 lbs – considerably more than the weight of most tractor drivers!

Ransomes

Ransomes, Sims & Jefferies of Ipswich, well known for their threshing machines, lawnmowers, ploughs and other farm equipment for two centuries, also made tractors from time to time since the early 1900s. James Ransome built a 20 hp four-cylinder petrol-engined tractor in 1903, which he demonstrated with a three-furrow plough, and about

ten years later Ransomes made a 35 hp oil-engined tractor that weighed about 10 tons and was 11 ft high to the top of the exhaust pipe.

The MG (Market Garden) cultivator, originally conceived in the early 1930s as a pedestrian-controlled garden tractor with Roadless Traction rubber-jointed tracks, was never put into production. However, the ride-on MG2 garden cultivator demonstrated to smallholders for the first time in 1936 was a success and the MG was made in its various forms for the next thirty years. It had a 6 hp Sturmey Archer air-cooled, single-cylinder side-valve engine with a dry sump. Lubricating oil, stored in a separate tank in the crankcase, was pumped to the bearings and a second pump returned it through a filter to the tank. A 4:1 reduction gear on the engine output shaft supplied power to a centrifugal clutch which engaged drive to the forward, neutral and reverse gearbox with a top speed of 2 mph in both directions. The gearbox had two inward-facing crown wheels and a central pinion. The direction of travel depended on which crown wheel was engaged with the pinion. The tractor moved off when the engine speed reached 500 rpm and it was steered with two hand levers connected to band brakes on the crown wheel shafts. The centrifugal clutch also served as an overload mechanism for the transmission, working on the principle that the engine speed would drop if the tractor was overloaded and the clutch automatically disengaged the drive. Both of the 6 in wide tracks were always under some degree of power which eliminated sliding and slewing when the tractor changed direction. Spacer blocks between the tracks and the chassis were used to obtain three track width settings.

A large number of MG2 crawlers were sold in France for vineyard work which probably explains why early sales leaflets gave specifications and capacities in imperial and metric units. Ransomes suggested to potential purchasers that 'the high-class baby track-type tractor is so simple that a boy could operate it and when ploughing the MG2 can turn so sharply that headlands will be practically non-existent.'

The MG5 with a 600 cc air-cooled petrol engine, still with a dry sump, replaced the MG2 in 1948. Like other petrol-engined tractors at the time it was possible to buy a conversion kit to run the engine on vaporising oil. The MG5 had the same gearbox and steering system as the MG2 but the power take-off speed was

257. The Ransomes MG2 cost £135 when it was introduced in 1936, the 400 rpm power take-off was an extra £1.10s.

258. The 15/25 hp Renault R3041 with a four-cylinder petrol engine was introduced to French farmers in 1946.

increased to 700 rpm. The instruction book pointed out that the MG5 was designed to do two-horse work at two-horse speed (2 mph) but Ransomes acknowledged the advance of mechanisation by increasing the top speed to 2½ mph. Distinguishing features of the MG5 included a petrol tank under the seat (it was on one side of the engine on the MG2) and a curved metal cover over the engine. A drawbar and a hand-lift toolbar were standard and an optional Neville hydraulic linkage attachment designed in Australia which cost £89.15s could be installed by the MG5 owner to provide fingertip control for toolbar equipment. Ransomes made a range of mounted and trailed implements for the MG including a toolbar with cultivator tines, ridging bodies, hoe blades and potato lifting body, a range of ploughs, disc harrows, a sprayer, and an earth scraper blade.

The Ransomes MG6 made its debut at the 1953 Royal Smithfield Show. Basically the same as its predecessors with a petrol or vaporising oil engine, the centrifugal clutch was retained but the reduction gears on the output shaft were replaced by a three forward and two reverse gearbox with a top speed of 4 mph. A hand-lift toolbar was standard and the optional hydraulic linkage with a power take-off driven pump added £52 to the price.

Industrial versions of the MG were announced in 1956. The Industrial Crawler Tractor (ITC) had the same track design with optional rubber blocks which could be bolted to the track plates to prevent damage when working on hard surfaces. The Industrial Tractor Wheeled (ITW) on pneumatic tyres had heavy roller chains linking both wheels on each side to give a solid four-wheel drive arrangement. The ITW was steered in the same way as the standard MG tractor using a similar system to that on modern skid-steer loaders.

The reign of the little blue crawlers was nearing its end when the MG40 was introduced in 1960. There was a choice of an 8 hp overhead valve two-stroke diesel engine or a side-valve petrol or vaporising oil engine with an Amal carburettor and a Wico magneto. The engines had a wet sump lubrication system with an oil pump and filter that started by hand. The diesel engine, which used about 3 pints of fuel per hour, had an ignition wick for cold starting. The MG40 also had a centrifugal clutch and the transmission consisted of a three forward and three reverse gearbox, a differential, spur gear reduction units and strengthened Roadless

rubber-jointed tracks. The track running gear was improved and wider track guards were added in 1962 when Ransomes offered an optional fibreglass bonnet and track guard extensions. The last Ransomes M40 crawler was made in 1966 but during a thirty-year production run more than 15,000 motor cultivators, including 3,000 MG2s, 5,000 MG5s, 5,000 MG6s and about 2,000 MG40s, were made in Ipswich.

Ransomes sold the agricultural machinery side of their business in 1987 but they re-entered the tractor market ten years later with six Japanese Shibaura four-wheel drive compact tractors with Ransomes green paintwork. The term 'compact' suggests tractors in the 10–30 hp bracket but the three- and four-cylinder diesel-engined Ransomes compacts ranged from the 18 hp CT318 to the 45 hp CT445. This tractor had five more horsepower than the New Fordson Major Diesel when it was launched in 1951! A mechanical or hydrostatic transmission was used on the 18, 20, and 25 hp tractors and the 33 hp CT333 HST had a hydrostatic drive with a shuttle-shift transmission. A creep speed box was standard on the 38 hp and 45 hp models. Textron acquired Ransomes in 1998 and Textron Groundcare distributed Iseki tractors from Ipswich.

Renault

Louis Renault started a car manufacturing business at Billancourt in France in 1898 and by 1913 10,000 cars had been made there. The first Renault tractor, based on the development of the Renault FT17 World War One battle tank, was made in 1919. The 30 hp Renault GP crawler tractor had a four-cylinder water-cooled petrol engine with the radiator at the back of the power unit and a three forward and one reverse gearbox. Renault built more than 400 GP crawlers in 1919 and 1920 and as a result of a big increase in the price of petrol in France some of these tractors were adapted to run on gas produced by a rear-mounted methane gas generator. The 20 hp Renault HO tractor, introduced in 1921, was a wheeled version of the H1 crawler with the same gearbox. The rear-mounted belt pulley could be set to run in a clockwise or anti-clockwise direction.

The 1926 Renault PE was the French company's first wheeled tractor with the option of steel wheels or solid rubber tyres. The radiator was behind the engine on early models, which was pleasant for the driver on frosty mornings. Renault moved the radiator to the front of the engine on the PE1 and later PE2 versions of the tractor. Landmarks in the ten-year production run of 1,840 Renault PE tractors included the introduction of hydraulic linkage in 1931 and optional electric starting and pneumatic tyres in 1933.

The 50–55 hp VI crawler, launched in 1932, was the first Renault tractor with a diesel or an optional 40 hp petrol engine with an electric starter. The VY 20–35 hp diesel wheeled tractor appeared in 1933 but a severe economic crisis in France in the 1930s resulted in the launch of the much cheaper Renault YL with an 8 or 15 hp petrol engine which was made between 1934 and 1938. The 10–20 hp Renault AFV made between 1938 and 1942 had a side-valve engine that ran on petrol or alcohol. An alternative gas-powered Renault AFVH was also made during the war years.

In an effort to rebuild the ailing French tractor industry, the state took over the Renault factory in 1945. The prefix letter R was added to Renault tractor model numbers during the period of state ownership when the organisation was known as the RNUR (Régie Nationale des Usines Renault). The 18–25 hp R304E and the gas-powered R304H with four forward gears and one reverse, pneumatic tyres and side-mounted pulley were made at Billancourt until 1948. The letter E indicated that the engine ran on 'essence' – the French for petrol. The R3040 launched in 1947 with a new dark orange colour scheme had a modified Renault 304 engine and optional electric lighting. The tractor was part of a major national effort to re-equip French farms and about twelve R3040 tractors were made every working day for the next two years.

The 18–25 hp R3041 first

introduced in 1948 was a restyled R3040 with a bonnet, side panels and radiator grille over the same engine. Made for four years the R3041 had a 627 rpm power take-off shaft, adjustable wheel tracks and a cable-operated lift for mounted implements. An optional hydraulic system was added in 1949. The R3042, with similar styling to the previous model, was made between 1949 and 1955 when there was a choice of a petrol, vaporising oil or alcohol engine for the standard and vineyard versions of the tractor and optional hydraulic linkage.

A total of almost 23,000 Renault R3042 tractors were made with a four-cylinder side-valve engine, magneto or coil ignition and an electric starter. Most had a petrol engine and eight out of ten tractors were sold with hydraulic linkage. It had four forward gears and

259. The Renault 3042 with a 22/30 hp petrol engine was launched in 1949 when it was advertised as more powerful than the previous model but just as economical to run.

one in reverse with a top speed of 13½ mph, independent foot brakes, a hand brake and a 552 rpm power take-off shaft was used for the optional bolt-on pump and hydraulic linkage. Other optional equipment included a side-mounted belt pulley and a 627 rpm power take-off shaft on the opposite side of the tractor for a side-mounted cutter bar mower. A sales leaflet described the R3042 as a '22–30 hp tractor which would run equally well on petrol, alcohol or vaporising oil' and purchasers were required to 'specify the fuel which will be employed' when ordering a tractor. The leaflet also provided examples of its use, including 'ploughing 5 acres of medium soil, 10 in deep with a double plough in one day'.

The Renault R7012 with a 32 hp three-cylinder Perkins engine and the R7022 with a 45 hp Hispano-Hercules diesel engine introduced in 1951 were based on the petrol-engined R3042 tractor. Diesel tractors were about 50 per cent more expensive than petrol tractors in France and many farmers still bought petrol-engined tractors in the early 1950s. The R3046 with an overhead-valve petrol engine superseded the R3402 in 1954.

A heavy one-piece cast-iron chassis had been used for Renault tractors since 1945 but unitary construction was adopted for the D (diesel) and E (petrol) series tractors launched in 1956. Renault opened a London office in the mid-1950s and the D series soon became popular with British farmers. However, fashions were at last changing in France and it was not long before demand fell for the Renault E30 with a 30 hp four-cylinder water-cooled petrol engine and the improved E31 that replaced it in 1957. The model numbers used for D and E series tractors denoted horsepower. The D22 had an air-cooled twin-cylinder MWM engine, a three-cylinder water-cooled Perkins was used for the D30 and the D35 had a three-cylinder air-cooled MWM power unit. The transmission was new with a two-stage clutch, six forward gears and one reverse or an optional twelve forward and two reverse gearbox and external drum brakes. The smaller D16 was added in 1958 to compete with the Massey-Harris Pony and Farmall Cub. It had a twin-cylinder MWM air-cooled engine but otherwise was similar to its bigger brothers. All of the D series tractors had the same hydraulic system as the earlier Renault R3042.

Engine speed was increased from 1,700 to 2,000 rpm

in 1960 when the four D series tractors with more horsepower and improved hydraulics became the N series. The Renault N73, N72, N71 and N70 model numbers seem to have little meaning as the smallest N73 was rated at 20 hp and the other models had 25, 35 and 40 hp diesel engines.

The Renault Super 5 and Super 7 launched in 1962 were the first of a new generation of D series tractors. The Super 7 had a 42 hp Perkins engine and the Super 5 had a three-cylinder 35 hp Renault water-cooled power unit that was the first diesel engine made by Renault since the late 1930s. Improved styling and more horsepower were the main changes on the 46 hp Super 7, 42 hp Super 6, 30 hp Super 3 and 25 hp Junior introduced in 1964.

In the hope of improving their sales most tractor makers launched new models at frequent intervals during the 1960s. New colour schemes, changed model numbers and the latest mechanical gimmicks were all used to tempt their farmer customers. Renault followed the trend with a new white radiator grille and the letter D for the D series was again added to the model number. The Renault Tracto-Control hydraulic system with draft and position control, implement float and optional double-acting spool valves appeared in 1965. A Selene front-wheel drive conversion kit for the Super 5D, Super 6D and Super 7D announced in 1966 was the last of the many changes made to the D series, which had been made since 1956.

Renault passed the 50 hp mark in 1963 when they launched the 55 hp 385 with a four-cylinder water-cooled Renault or air-cooled MWM engine. The tractor was well advanced for its time with five forward gears and one reverse, a 540/1,000 rpm and ground speed power take-off. Other features included two engine-mounted hydraulic pumps, one provided oil for the three-point linkage, spool valves and hydraulic disc brakes, the second serviced the power steering system. Earlier problems with leaks and burst pipes in the 385's high-pressure hydraulic circuit were solved with a model change and improved safety features in 1964. The Renault 385 with an MWM diesel engine became the 385 Master 1 and the Renault-engined version was known as the 385 Master 2. Both tractors were discontinued in 1969.

Renault tractor production was moved to a new factory at Le Mans in 1967 and the 55 hp Renault 86 and 88 with the same engines as the earlier 385 Master tractors were launched in 1968. The Renault 86 had an MWM engine and a Renault power unit was used for the 88 but the tractors were quite different to the 385 with more angular styling, a twin-range four forward and one reverse gearbox and a less complicated hydraulic system.

A few four-wheel drive Renault 456 tractors with a 42 hp three-cylinder MWM engine and similar styling to the Renault 86 and 88 were made between 1968 and 1971. A Selene front-wheel drive unit was used at first but later models had a Zetor front axle with a central drive shaft.

The 30–46 hp Renault 53, 55, 56 and 57 diesel tractors and the 45 hp petrol-engined Renault 58 replaced the Super D series in 1968 but the days of petrol-engined tractors were at last coming to an end and the Renault 58 was withdrawn in 1969. The first Renault VF narrow orchard and vineyard tractors were made in 1968. The four models included the 33 hp Renault 50 Vineyard tractor which was only 890 mm wide. About 18,000 VF orchard tractors had been made when it was discontinued in 1977.

The Renault tractor operation was given a separate identity in 1969 when it became Renault Motoculture. The 70 hp air-cooled Renault 94 and 80 hp water-cooled Renault 96 with MWM engines, announced in 1969 to meet the growing demand for more power,

were made until 1973. Apart from a hand clutch lever for the 540/1,000 rpm and ground speed power take-off the Renault 94 and 96 were mechanically similar to the competition with power steering and hydraulic multi-plate disc brakes. There was a choice of a cab or roll bar and the Super Comfort driver's platform had four rubber mounting blocks to isolate it from transmission vibration.

Renault also met a local demand for medium-powered four-wheel drive tractors in 1972 when they sold the Italian-built 34, 45 and 57 hp Carraro tractors in Renault livery as the 321-4, 451-4 and 571-4.

More changes in the early 1970s included an improved Tracto-Control hydraulic system, power steering became standard across the range and an optional trailer braking system was introduced to meet new road vehicle regulations in France. New models with 51–90 hp MWM engines and optional four-wheel drive were added 1971 to meet the demand for still more power. The prefix '4' indicated that the tractor had either a ZF or a Zetor front axle and the four-wheel drive Renault 4-98 with a six-cylinder 90 hp engine and hydrostatic steering was the most powerful Renault tractor in the early 1970s.

The last Renault tractors with Renault diesel engines were launched in 1973 at the SIMA agricultural show in Paris. Within a couple of years the Renault range included models with 30–145 hp MWM engines. The

260. The 1970s Renault 651 with a 65 hp four-cylinder engine had twelve forward and three reverse gears provided by a synchromesh main gearbox and a three-speed range box.

model number indicated engine horsepower, the figure 4 denoted four-wheel drive and the letter S was used for tractors with a cab. For example, the Renault 651-4 S was a 65 hp four-wheel drive tractor with a cab.

The Renault 301, 461, 551 and 651 were launched in 1973, all with MWM engines. The two-wheel drive 301 with a twin-cylinder air-cooled engine and a six forward and reverse or ten forward and two reverse gearbox was the only model without hydrostatic steering. The 461, 551 and 651 had three- or four-cylinder air-cooled engines, a three-range four-speed gearbox with a shuttle reverser giving twelve forward and reverse gears and optional Carraro front-wheel drive.

The 751/751-4, 851/851-4, 951/951-4 and 1151-4 appeared in 1974 and the six-cylinder turbocharged 1451-4 completed the range in 1975. The flagship 1451-4 had fifteen forward and six reverse gears, power steering and a safety cab with optional heating and air conditioning on the anti-vibration driving platform. The Renault Automatic Blocamatic progressive diff-lock for the front axle was introduced in 1977 and with extra power for the high horsepower models they were renumbered as the 781, 891, 981 and 1181. Some of these tractors were only made for three or four years but the 461, 551 and 651 survived until 1986. The 781 was discontinued in 1987 and the 891/891-4 remained in production until 1989.

Apart from a brief presence in the UK in the late 1950s there had been no official Renault Motoculture sales outlet in Britain since the first Renault HO wheeled tractor appeared in 1921. A significant event for Renault in 1977 was the appointment of their first UK dealer at Thorne near Doncaster and the French company established their own UK premises in the Doncaster area in 1978. Renault Motoculture became Renault Agriculture in 1980 with a separate logo to distinguish tractor dealers from car and truck distributors. The company moved to Shipston-on-Stour in Warwickshire in 1981.

The new 'T' cab, which could be lifted clear of the transmission housing in less than an hour for repair work, was described as 'simply years ahead of its time' when the Renault TX series was launched in 1981. Three models with 103, 123 and 135 hp six-cylinder air-cooled engines had a twelve forward and reverse synchromesh gearbox in three ranges, an optional creeper gearbox and hydraulic dry twin-disc brakes.

The re-styled TX series tractors were the first Renaults with optional front linkage and a new light orange colour scheme. The medium-powered TS range was launched in 1981 with 84 and 93 hp models added to the TX series in 1982. The 84 hp TX95, for example, had an independent 540/1,000 rpm power take-off, a forward/reverse shuttle lever for loader work and a tip-up cab.

A computer, programmed to help conserve fuel and achieve maximum productivity, was a new feature in the TX and TS cabs from 1983. Renault's Ecocontrol system recorded engine speed and fuel consumption was measured by a sensor in the exhaust manifold. Two needles on the Ecocontrol gauge displayed this information and the engine was running at maximum efficiency when both needles were in the green zone. A low-line RS cab for tractors up to 93 hp was introduced in 1984 for farms where the buildings were too low for the standard TS cab. Renault followed the example of some other tractor makers between 1984 and 1986 when they sold a range of Japanese Mitsubishi compact models in Renault livery to farmers and growers in France.

The Renault LS tractors and the Perkins-engined SP models with improved hydraulic pumps, twin assistor rams and new cabs appeared in 1986. The Renault ACET and TCE electronic systems for their high-powered models were introduced in the same year. The ACET system (Aid to Economical Tractor driving) was an improved Ecocontrol unit with a screen displaying fuel consumption and indicating the most economical gear ratio and engine speed for the work in hand. TCE (Tractor Control Electronique) was the Renault version of electronic hydraulic linkage control with remote external switches on the rear mudguards included in the package. The launch of the TX16 tractors with a sixteen forward and reverse transmission and a new range of fruit and vineyard tractors made by Carraro completed a busy year for Renault. The Hydrostable cab introduced in 1987 was used for tractors from the TX110 upwards. The cab, with independent shock absorbers and anti-surge bars to isolate the driver from vibration, had its origins in the SuperComfort driver's platform mounted on four rubber pads first seen in 1969.

The 83 hp 90-34M and 90-34T tractor models with a standard specification and no optional extras

261. The six-cylinder 123 hp Renault 133-54TX cost £41,190 in 1990.

appeared in France in 1988. The 70-34PX and 133-54TX were added in 1991. Launched in 1989, the top of the range four-wheel drive 145 hp Renault 155-54 with a six-cylinder MWM engine and a sixteen forward and reverse transmission was available with the optional TZ Hydrostable cab or standard TX cab.

Improved electronics led to the launch in 1989 of the Tractoradar system which used a radar sensor to monitor forward speed, calculate wheelslip, distance travelled and work done. The system could also be programmed to automatically reduce the working depth of a mounted implement when wheelslip exceeded a pre-set level.

The 165 hp 175-54TZ was launched in 1990 to satisfy the continuing demand for still more horsepower but less than 200 were made during its two-year production run. The tractor had a turbocharged six-cylinder MWM engine, a twenty-four forward and eight reverse speed Steyr transmission and ACET computer system. The twenty-four forward and eight reverse Tractoshift transmission with three change-under-load speeds engaged with hydraulic clutch packs in each forward gear was an added option for the TZ models from 1991.

Renault launched the 170 hp 180-94TZ Multi-shift and re-introduced the 100 hp six-cylinder 106-54TL lightweight tractor to UK farmers in 1992. The Multi-shift 180-94TZ transmission with twenty-seven forward and reverse gears was made by SAME-Lamborghini-Hürlimann. A 150 hp TZ Multi-shift model was added in 1993. Multi-shift tractors were not very popular but they remained in production until the Ares 700 range was launched in 1997. The two- and four-wheel drive air-cooled 55 hp Ares introduced for the smaller farm in 1992 had a twelve forward and reverse gearbox with a top speed of 17 mph.

The distinctive gold-painted Renault 120-54 and 155-54TZ Nectra with 145 hp MWM engines appeared in 1994 and the Tractorshift transmission, improved in the same year with clutchless direction changes, was re-named the Tractronic transmission.

The 54–83 hp Renault Ceres tractors with sloping bonnets and two cab options, named after the Greek goddess of agriculture, were launched in 1993. The MWM engines and back axles were used for the previous range but the Ceres tractors had a new ten-speed synchromesh gearbox with a reverse shuttle. Optional extras included a creep speed box and the Twinshift change-on-the-move transmission with an

262. The Ares 725 RZ – one of sixteen Renault models launched in 1999 – had a 165 hp constant power engine, Revershift transmission with fingertip clutchless changes of direction and optional load sensing hydraulics. The smallest 89 hp Ares 540 RX cost £29,500 and the most powerful 194 hp Ares 735 RZ was £56,860. (Renault)

electro-hydraulic splitter. Renault changed to John Deere engines in 1995 in preparation for new exhaust emission regulations and some Ceres tractors were painted green and yellow and badged as the John Deere 3000 series.

Cergos and Ares tractors made their debut in 1997. The 75, 85 and 95 hp Cergos models had a similar transmission system to the Ceres with optional Twinshift and creep speed gearboxes. The four-cylinder 500 series and the six-cylinder 600 and 700 series Ares were launched in eight power sizes from 85–165 hp. A sixteen forward and reverse gearbox was standard on the 85, 95 and 100 hp tractors while the more powerful models had a thirty-two speed transmission. Three Fructus fruit tractors, three Ceres models and the 106-54TL and 155-54TX completed the Renault range on the UK market in 1997.

An improved three-model Herdsman range with 52–76 hp Deutz air-cooled engines and a new rounded bonnet, seen for the first time at the 1998 Royal Smithfield Show, had an adjustable steering column and an electronic monitoring system for the hydraulic linkage.

The Renault Herdsman, Fructus, Ceres, Cergos, Ares and High Range tractors were current in 1999. The two- and four-wheel drive 65 hp Ceres models had a ten forward and reverse gearbox while the 75 and 85 hp Ceres and Cergos tractors had a twenty forward and reverse transmission. The nine four-wheel drive Renault Ares tractors with Hydrostable full-suspension cabs included the 85, 95 and 100 hp models with sixteen forward and reverse gears and eight models in the 110–185 hp bracket had a thirty-two forward and reverse gearbox.

263. The late 1990s Ares 815 RZ and 825 RZ were the first Renault tractors with a full powershift transmission. (Renault)

The new four-wheel drive 52–76 hp Pales, 102–154 hp Temis and 197–250 hp Atles were included in the 2001 price list and Renault introduced the 90–205 hp Ares II 500, 600, 800 and 900 ranges to UK farmers at the 2002 Royal Show. The 90–110 hp 500 series had four-cylinder engines while the 600 and 800 series had six-cylinder engines. With the exception of the turbocharged and intercooled 175 hp 826 which had a full powershift transmission, a thirty-two forward and reverse Quadrishift II was standard across the range. Other features of the new Ares II range included the Renault Quadractiv load-sensing splitter used to select power or Eco working mode to give the most economical engine speed and gear ratio.

Four new Renault Celtis models with 75–102 hp John Deere DPS Powertech Tier II engines under their sloping bonnets were added to the existing Fructus, Pales, Temis, Ares II and Atles tractor ranges in 2003. The Celtis, with a twenty forward and reverse transmission, had a car-type heating and ventilation system in the cab.

Claas acquired a 51 per cent in Renault Agriculture in 2003 and the first tractors in the green Claas livery were sold in Europe in the same year. The Claas Celtis, Ares, Atles, Fructus, Nectis and Pales ranges along with the latest 335 hp Xerion 3300 were introduced to British farmers at the 2004 Royal Smithfield Show.

Roadless

Lt Col Philip Johnson, who founded Roadless Traction Ltd at Hounslow in Middlesex in 1919, spent World War One developing battle tanks. He used this experience to design tracked conversions for steam wagons and commercial vehicles. The first Roadless involvement with farm tractors was in 1922 and the Roadless Orolo track units with two or three bottom rollers, introduced in 1926, were self-contained track units made in various sizes. Roadless made full-track conversions for the Fordson Model N, Garrett, Case and other tractors in the early 1930s using rubber-jointed tracks.

264. A number of Fordson Model Ns with Roadless half-track conversions were sold to the Air Ministry during World War Two.

Sales literature explained that the Garrett Roadless diesel tractor with Roadless rubber-jointed tracks could be supplied with either a Blackstone or Aveling diesel engine, rated at 24–30 drawbar hp, a single-plate transmission clutch and a three forward and reverse gearbox. Roadless also developed rubber-jointed tracks for Bristol and Ransomes MG crawler tractors.

Some users, especially on tractors used for military purposes with a front-mounted winch or crane, experienced problems in the early days when tractors on Roadless full tracks pitched forward when working at speed. Roadless introduced a wheeled fore-carriage Traction in 1938 to overcome the problem.

Roadless Traction spent much of World War Two making full-track and half-track conversions for Fordson and Case tractors with many of them sold to the Air Ministry. Roadless engineers were also busy developing a new DG (Driven Girder) half-track conversion for the Fordson Model N which appeared in 1944. The large footprint area of the DG track, equivalent to that of a 20 ft diameter wheel, considerably increased tractive efficiency compared with a wheeled tractor. Roadless DG half-tracks were available for the Fordson E27N Major shortly after its introduction in 1945 and Roadless Traction made half-track conversions for Case, David Brown, Fordson, Field Marshall, Massey-Harris, Nuffield and other popular tractors well into the 1950s.

Roadless returned to crawler conversions in 1950 with the Roadless Model E full track based on an E27N Major with a Fordson vaporising oil or Perkins P6 diesel engine. When the Fordson E1A Major superseded the E27N in 1951 Roadless designed and built their own prototype crawler which they exhibited at the 1953 Royal Show. However, the tractor was not put into production and it was back to the drawing board. The first Roadless J17 crawlers, based on the Fordson E1A Major, were made at Hounslow in 1954. The J17 had an improved rubber-jointed track which was also used for later Roadless crawlers based on Fordson Power Major and Super Major skid units.

Crawler sales fell away as demand grew for the

265. Introduced in 1945 Roadless DG half tracks were made for the E27N Fordson Major and other popular tractors of the day.

266. Manuel
front-wheel drive
conversion kits
were made for
the E27N
Fordson Major.

267. The drive shaft for the Manuel front-wheel drive conversion on the Fordson Major ran from a transfer box behind the main gearbox to the front axle differential housing.

more versatile four-wheel drive tractors. Roadless Traction obtained a licence from Selene in Italy to make the Manuel front-wheel drive conversion kits at Hounslow for the Fordson Major. The Manuel-Roadless four-wheel drive conversion, introduced in 1956, had a transfer box sandwiched against the gearbox, a drive shaft, front axle differential and front wheel drive hubs. Four-wheel drive soon became popular with farmers, a similar Roadless conversion for the Fordson Dexta appeared in 1959 and conversions for the Major and Dexta were made until 1964. It was also possible to buy a four-wheel drive conversion kit for existing tractors at a cost of about £500 plus a £25 fitting charge. Roadless Traction introduced a four-wheel drive conversion for the International Harvester B450 in 1963 which they made for seven years and similar conversion kits were available for the International Harvester 614, 634 and 414. Robert Eden Ltd of London marketed Manuel four-wheel drive conversions for Massey Ferguson 35 and 65 tractors.

The green and yellow six-cylinder Roadless Ploughmaster 6/4 was the first of many Ploughmaster tractors made during the next twenty years. The green and yellow colour scheme, which caused some confusion with John Deere tractors, was soon changed to blue. New Roadless models were inevitably linked to changes made by the Ford Motor Co. The Ploughmaster 65 and the Ploughmaster 90, announced in 1965, were based on Ford 5000 skid units with optional Select-O-Speed. The six-cylinder Ford engine rated at 90 hp was used for the Roadless 90 but within a year Ford introduced a new six-cylinder power unit and the tractor became the Ploughmaster 95.

Roadless moved down the power range in 1966 with the Ploughmaster 46 which they made in small numbers for about six years. The tractor was basically a 46 hp Ford 3000 with a Selene front-wheel drive conversion.

The 80 hp Ploughmaster 80, introduced in 1967, was a turbocharged version of the Ploughmaster 65 with a

268. The Ploughmaster 6/4 had a six-cylinder Ford industrial engine.

269. Introduced in 1968 the 115 was the first four-wheel drive Roadless tractor with equal-sized wheels.

270. The Ploughmaster 75 was based on a Ford Force 5000 skid unit. It replaced the Ploughmaster 65 in 1968.

bigger engine radiator, an oil cooler and an assistor ram to increase the lift capacity of the hydraulic linkage. The tractor, which cost £2,165, was not popular and was withdrawn in 1968. The Ploughmaster 75, based on the improved 75 hp Ford 5000, was launched the same year. The four-wheel drive Roadless 115 with equal-sized wheels was introduced at that year's Royal Show. The earlier prototype 115, disguised with red paint, was dubbed the Red Devil. The Roadless 115 had many Ford-sourced components including a strengthened gearbox, heavy-duty clutch, hydrostatic steering and a 115 gross hp six-cylinder Ford industrial engine under its bonnet. To meet the new farm safety regulations in 1970 Roadless tractors were supplied with Duncan safety cabs.

Ford introduced a new and more powerful range of industrial engines in 1971, which proved to be a busy year at the Hounslow factory. The Roadless 120 and 94T tractors were launched in 1971, a new engine was used for the Roadless 115, the Ploughmaster 95 and the Roadless 95 with a planetary axle. The Roadless 120 was a 120 gross hp version of the 115 and the turbocharged Roadless 94T, based on a 94 hp Ford 7000 skid unit, had a new front axle with a diff-lock and planetary reduction gears. The 104 hp Roadless 105 with unequal-sized wheels and a planetary front axle replaced the Roadless 95 in 1974.

The launch of the Ford 600 series in 1975 heralded the arrival of the last new Ploughmaster when Roadless introduced the 98 and the Ploughmaster 78. The Ploughmaster 78, based on a Ford 6600 skid unit, superseded the popular Ploughmaster 75 and a Ford 7600 turbocharged engine and transmission was used for the Roadless 98. The Roadless 118, which replaced the Roadless 105 in 1976, had a 114 hp Ford engine, a Ford 7600 transmission, hydrostatic steering and a Lambourn safety cab.

Most major tractor manufacturers were making their own four-wheel drive models in the late 1970s and time was running out for Roadless and their competitors. The Ploughmaster 78 and the 98 together with the Roadless 118 and 120 were still being made in 1980 when Roadless Traction Ltd moved to Sawbridgeworth in Hertfordshire. The partial-axle Roadless 780 and 980, based on Ford 6600 and turbocharged 7600 skid units, were launched in 1979 with a new front axle design and an improved turning circle. They were the last new Roadless models but very few were sold so Roadless turned their attention to specialist articulated tractors, including the hydrostatic drive Logmaster for forestry work.

Roadless Traction made two- and four-wheel drive Teleshift mechanical handlers with telescopic booms at Sawbridgeworth for RWC at Ledbury in 1982 but

271. The JJ Thomas Ninety Five-100 conversion of a late 1960s Ford 5000 completed in 1980 has a 114 hp six-cylinder Ford engine.

went into liquidation in 1983 and LF Jewell Ltd bought the Roadless business. The new owners, who were Ford tractor dealers in Somerset, bought the stock of spare parts and built a few equal-sized four-wheel drive Jewelltrac 103 and 120 tractors. Based on original Roadless designs the 103, which cost £25,000, had a 103 hp turbocharged Ford 7610 engine and a constant mesh gearbox. A Ford 7610 skid unit with a 125 hp industrial six-cylinder engine was used for the Jewelltrac 120. Dual Power, a sixteen forward and four reverse gearbox and two-speed power take-off were standard on both models. LF Jewell went into liquidation in 1991.

JJ Thomas Tractors Ltd, a Ford dealer at Banbury in Oxfordshire, used refurbished second-hand Ford 5000 tractors and some Roadless parts, including a heavy-duty clutch, a modified front

axle bracket and a longer bonnet, to build the Thomas Ninety Five-100 tractors. Many of these early 1970s Ninety Five-100 tractors with a six-cylinder 104 hp or 114 hp Ford engine and an eight forward speed gearbox were sold in America.

When JJ Thomas Tractors introduced the Farmair three-line air braking system for trailers and slurry tankers in 1978 they were making six-cylinder Mk II and Mk III Ninety Five-100 tractors. The Mk II was based on a Ford 7600 skid unit and the Mk III was a stretched version of the Ford 7700. A four-wheel drive version of the Mk III with a Schindler front axle appeared in 1979 when the company was still fitting Ninety Five-100 conversion kits to farm-owned Ford 5000, 7000, 7600 and 7700 tractors. The JJ Thomas Ninety Five-120 with a flat floor cab was introduced in 1980 and made in limited numbers for the next two or three years.

Russell

Russells of Kirbymoorside in Yorkshire made the Russell 3-D self-propelled tool carrier, which was similar to the David Brown 2D, in the mid-1970s. The 3-D had a rear-mounted twin-cylinder Lombardini air-cooled 20 hp diesel engine with an electric starter. A single lever controlled the infinitely variable transmission with a hydrostatic motor in both rear wheels, which were supplied with oil by a variable displacement pump. Another lever controlled the hydrostatic transmission with a maximum forward and reverse speed of 9 mph and levers were also used to raise and lower the mid-mounted toolbar and the optional rear wheel mark eradicator tines. The tractor had independent disc brakes and a parking brake and the wheel track was adjustable from 48–76 in. Attachments for the mid-mounted toolbar included cultivator tines, hoe blades and seeder units. The Russell 3-D, which cost about £4,000 in 1978, was still available in 1986.

A sales leaflet explained that the excellent operator vision from the driving platform with an almost unobstructed view of the rows allowed the driver to place the tines or blades as close as possible to the growing plants with complete confidence. There was easy access to the padded seat on the driving platform and the controls including the hydraulic levers and throttle were all within easy reach of the driver.

272. The Russell 3-D self-propelled toolbar had a twin-cylinder air-cooled diesel engine.

Chapter 8

SAME – TYM

SAME

Francesco Cassani made his first diesel-powered tractor with a twin-cylinder water-cooled engine in 1927 and in the mid-1930s Italian farmers were using 40 hp Cassani tractors with similar twin-cylinder engines. Cassani had also designed and built engines for aircraft, boats and lorries before he established a company called Società Accomandita Motori Endotermici in 1942. The initial letters form SAME (pronounced Sammy) and this name was used in 1948 for a self-propelled, tricycle-wheeled ride-on cutter bar mower with a single-cylinder air-cooled petrol engine. The mowing machine, steered by its single castor wheel, was also used for ploughing, cultivating, pumping water, driving an electricity generator and other work.

The conventional SAME 4R/20 tractor appeared in 1948 and the four-wheel drive DA25DT (DT added to the model number indicated four-wheel drive) introduced in 1952 was the first Cassani tractor with an air-cooled full diesel engine. Several models of SAME tractors were made in the 1950s including two- and four-wheel drive versions of the DA25 and DA38, the three-cylinder DA47 Supercassani and the 21 hp Sametto orchard tractor. The Sametto had a single-cylinder air-cooled diesel engine, a creep speed gearbox, diff-lock, a four-speed power take-off with ground speed and optional four-wheel drive.

The twin-cylinder 340, introduced in 1958, was the first SAME tractor with the new Cassani Automatic

273. Two- and four-wheel drive versions of the 62 hp three-cylinder SAME 360 had a direct injection diesel engine and a five forward and one reverse gearbox. Sales literature described the endometer (tractor meter) as 'an absolute novelty, the magic eye of your tractor'.

274. The early 1960s two- and four-wheel drive Sametto 120 had a single-cylinder air-cooled 21 hp engine, six forward gears and one in reverse with a top speed of 10 mph and four power take-off shafts.

Linkage Control system with lower link sensing. Within a couple of years the linkage was also optional for the 42 and 62 hp SAME 240 and 360 tractors. A sales leaflet for the SAME 240 described the automatic linkage control as a system which 'keeps the working depth unchanged on all implements including half-carried types as well as drawn ones which no other device can achieve'. It was also explained that the 240, with a windshield and a hydraulic shock absorber for the seat, was comfortable to drive and the new endometer (tractor meter) an 'absolute novelty – the magic indicator of your tractor'.

The 82 hp 480 with a four-cylinder air-cooled diesel engine and five forward gears and one reverse with a top speed of 15½ mph was the most powerful SAME tractor in 1960. It had a hand and a foot throttle, the power take-off shaft turned at 690 rpm when the engine was running at 1,700 rpm and the ground speed related power take-off for power-driven trailers,

275. The windshield on the 480 was a feature of SAME tractors in the late 1950s and early 1960s.

276. The SAME Leone has a central clutch-less drive shaft to the front axle. Some of these tractors had a downswept exhaust; others had twin vertical exhaust pipes.

seeders and planters. Optional extras included four-wheel drive and a belt pulley attachment for the power take-off.

SAME tractors were not available in Britain until 1964 when Cornish and Lloyds of Bury St Edmunds imported the Leone and Centauro to meet East Anglian farmers' need for a tractor which had a hydraulic system with enough capacity to lift a heavy reversible plough. The 67 hp Leone 70 with an air-cooled V4 direct injection engine and twin vertical exhaust pipes had eight forward and four reverse gears and disc brakes. The tractor with a set of front-end weights cost £1,995 in 1968 and the 55 hp Centauro with a V4 diesel engine and a similar gearbox was £1,755.

AC Bamlett Ltd at Thirsk in Yorkshire became the sole UK importer for SAME tractors in 1969 and this relationship continued in the early 1970s when SAME had an annual production of around 17,000 tractors. There was further growth in 1972 when Lamborghini joined the SAME group. The Lamborghini tractors retained their separate identity and Sametrac, also based at Thirsk, were the UK concessionaires for the tractors.

The 1974 SAME tractor range included the four-wheel drive 56 hp Minitauro 60, the 78 hp Saturno 80 and the 98 hp Drago with three-, four- and six-cylinder air-cooled engines respectively, a dual clutch and an eight forward and four reverse speed gearbox. There were five models in the 56–126 hp bracket when the Tiger and Buffalo with flat floor cabs appeared at the 1976 Royal Smithfield Show. The 126 hp six-cylinder Buffalo, which cost about £15,000, had telescopic ends on the lower lift arms and dual-speed independent power take-off. The five-cylinder 100 hp Tiger with hydrostatic steering, dual-speed power shaft and a price tag of about £13,000 was advertised as 'the big five with a hundred smooth horses'.

The Swiss-based tractor maker Hürlimann joined the group in 1977 and the name SAME-Lamborghini-Hürlimann was adopted in 1978. The 82 hp Leopard 85E (Export) with an air/oil-cooled engine, a twelve forward and three reverse synchromesh gearbox, hydrostatic steering and oil-immersed brakes appeared in 1979. Alternative twenty-four forward and six reverse or twenty forward and five reverse transmissions could be specified for the Leopard 85E.

277. The driving platform on the SAME Leopard 85, with an 82 hp four-cylinder air-cooled engine, was mounted on rubber suspension blocks.

SAME (UK) Ltd took control of marketing SAME, Lamborghini and Hürlimann tractors in the UK in the late 1970s. SAME tractors on sale in the UK in 1979 included two- and four-wheel drive versions of the Taurus 60, Centurion 75, Leopard 85, Panther 90, Tiger 100 and Buffalo 130. The figures after the model name indicated the approximate engine horsepower. The specification for the Taurus, which cost £7,325, included a fourteen-speed gearbox, hydraulic lower link sensing, hydrostatic steering, a two-speed power take-off and a de luxe quiet cab. An advertisement explained that there were no optional extras for the Taurus with its miserly fuel consumption, reliability and driver comfort.

The four-wheel drive Hercules 160, the largest tractor yet made by SAME, was introduced to British farmers in 1981 along with the Centauro 70 and Trident 130. The turbocharged and intercooled Hercules 160, which cost £26,545, had a flat deck cab

with the added luxury of tinted glass, air conditioning and a radio. Other features included lower link sensing hydraulics with SAME's Automatic Control Unit (ACU), separate 540 and 1,000 rpm power take-off shafts and a diff-lock that remained engaged while travelling in a straight line and automatically disengaged when the tractor changed direction. The two- and four-wheel drive Centauro 70 and Centurion 75 with air-cooled 68 hp four-cylinder engines had a twelve forward and reverse gearbox, lower link sensing and hydraulic brakes. The 125 hp four-wheel drive Trident 130, a replacement for the Buffalo, had a twelve-speed gearbox, epicyclic reduction gears on both axles and the very latest cab with air conditioning, tinted glass and an adjustable steering column.

The turbocharged 88 hp Leopard 90 joined the Leopard 85 in 1982, a year when SAME made more than 25,000 tractors and held sixth place in the manufacturer's league table. The Galaxy and the Laser

110, 130 and 150 models with a new orange, grey and white colour scheme were launched in 1983. The 165 hp Galaxy, which replaced the Hercules, and the new six-cylinder 110, 125 and 145 hp Lasers had twenty-four forward and twelve reverse synchromesh gearboxes, open centre hydraulics with lower link sensing and a fully air-conditioned cab.

New models and the addition of some Lamborghini features on SAME tractors were the main changes in 1984. Tractors over 60 hp had pedal-operated hydraulic trailer brakes and oil-immersed front disc brakes became standard on models over 110 hp. The 1984 UK price list included the two- and four-wheel drive Condor 55, the Minitaurus 60 fruit tractor, the Taurus 60, the Centauro 70, the Centurion 75, the Leopard 90 and the Jaguar 100. The S110 and the four-wheel drive S130, S150 and S170 completed the list. The turbocharged four-cylinder 88 hp S90 and five-cylinder 98 hp S100 with twenty-four forward and twelve reverse speeds and open centre, lower link sensing hydraulics were added later in the year.

SAME (UK) Ltd moved to Barby near Rugby in 1986, when in common with other manufacturers, about 80 per cent of their tractors were sold with four-wheel drive. With the move completed the company became SAME-Lamborghini (UK) Ltd and a range of water-cooled Lamborghini tractors was introduced to British farmers. The main differences between the two makes were their colour schemes and engine cooling systems. 'Air neither freezes or boils' was a SAME advertising slogan used at the time for their air/oil-cooled engines with an oil spray system to cool the underside of the pistons. The 88 hp SX90T with a turbocharged air-cooled engine in both two- and four-wheel drive format was added to the SAME range in 1986. The 874-90 with the exception of its water-cooled engine was the equivalent tractor in Lamborghini livery.

The SAME Synchro-power electro-hydraulic control system replaced the pedals previously used to operate the diff-lock, four-wheel drive and four-speed economy power take-off on their medium-power tractors including the 60–90 hp Explorer II range introduced at the 1988 Royal Smithfield Show. The 'comfortable, climatised operational station' (cab) on the six-cylinder Antares 110 and 130 launched in 1990 had the very latest computerised controls including an

278. A twenty-seven speed electronic transmission was standard on the SAME Titan.

electronic engine speed regulator (governor), radar-operated wheelslip control and a performance monitor. A reverse-drive version of the 70 hp SAME Aster, also known as the Dual Trac 70, with front linkage, power take-off and various modifications to the driving controls, also appeared in 1990.

The Explorer II models, with a twelve-speed synchromesh reverse shuttle gearbox and low-profile cab for stock farmers, and the 153 hp Titan 160 were launched in 1991. The Titan 160's electronically controlled air-cooled engine and powershift transmission, hydraulic linkage and front-wheel drive engagement were also used for the new 160 hp Lamborghini Racing and 132 hp Hürlimann H-6135 Elite tractors.

The 1992 Fruetto II orchard tractor took electronic controls a stage further. The 75 hp four-cylinder air/oil-cooled diesel tractor had a computerised engine management system instead of the usual mechanical governor to give more accurate control of engine speed for spraying and similar work. The Solaris 25, 35 and 40 four-wheel drive compact tractors

made by Zetor for SAME were also marketed in Lamborghini livery. The Solaris range had 25, 33 and 40 hp Mitsubishi engines with an anti-start security device to deter thieves, a twelve-speed synchromesh gearbox with a reverse shuttle lever on the steering column, load-sensing power steering and a three-speed independent power take-off.

More activity in 1994 saw the introduction of the more powerful 159 and 189 hp Titan 160 and 190 with a twenty-seven speed electronic powershift transmission and the 110 and 130 hp Antares with a thirty-six speed shuttle transmission. The Mk II electronic engine management system which the driver could use to pre-programme engine speed for working and headland cycles was a new feature on the Antares tractors. The 80–100 hp SAME Silver models with the choice of electronic or manual controls for livestock farms made their debut at the 1994 Royal Smithfield Show. An advertisement explained that the new 21st century cab gave easy access to a comfortable working environment and the optional bionic arm will turn the driver into a pilot.

279. Five 135–193 hp models of the SAME Rubin with a twenty-seven forward and reverse shuttle transmission were current in 2001.

The SAME-Lamborghini-Hürlimann group acquired Deutz-Fahr in 1995. Watveare Ltd at Westbury in Wiltshire remained the UK distributor for Deutz tractors until 1997 when SAME Deutz-Fahr (UK) established Deutz-Fahr (GB) as a separate company but also based at Barby near Rugby. Other developments in that year included the introduction of a hydrostatic transmission for the Solaris, an improved transmission for the Silver range and optional automatic APS transmission for the Titan and Antares.

The Titan 145, 160 and 190 APS automatic transmission continually adjusted tractor speed to suit soil conditions and the Agroshift system on the Antares II enabled the driver to make three clutchless changes in each gear to give a total of fifty-four forward and reverse speeds. The Directronic electro-hydraulic control for the Silver range transmission had two multi-plate hydraulic clutch packs to simplify repetitive shuttle operation. One clutch pack controlled forward travel while the other provided reverse drive through an epicyclic unit. A built-in microprocessor carried out a diagnostic check of the Directronic system every time the driver started the engine.

To celebrate seventy years of tractor production SAME launched the Golden 60 and 75 hp vineyard and compact tractors in 1997. About sixty new SAME Deutz-Fahr models appeared over the following eighteen months and the SAME Rubin 120, 135 and 150 were launched at the 1998 Royal Smithfield Show. The model numbers still indicated the horsepower of the six-cylinder liquid-cooled engines. Other features shared with three new Lamborghini Champion models included an oil-immersed multi-disc clutch, an eighteen forward and reverse powershift transmission and a cab with a car-like interior.

SAME tractors on the UK market in 1999 included the Dorado, Explorer, Explorer Top, Rubin and Silver together with the Antares II and Titan which were discontinued later in the year. The 60 and 70 hp Dorado with electronic engine speed control and a twenty forward and ten reverse Agroshift transmission were discontinued in 2004. The Italian-styled 70, 80 and 90 hp Explorer Top, which was made until 2003, had a triple engine cooling system,

hydraulically operated clutch and disc brakes and a synchromesh shuttle gearbox with the option of twenty-four forward and twelve reverse speeds or forty forward and reverse gears. The improved 80–130 hp four-wheel drive SAME Silver models superseded the earlier models in 2000. Electronic hydraulic linkage control, an engine management system and a power shuttle transmission were optional extras for the 80–100 hp Silver models with a sixty forward and reverse transmission. The six-cylinder Silver 110 and 130 had a fifty-four forward and reverse gearbox. The SAME price list in 2002 also included the 55 and 60 hp Rock crawler tractors with three-cylinder engines and the choice of an eight forward and reverse or a twelve forward and eight reverse transmission. The Rock 55 had narrow steel tracks and the 60 hp model ran on rubber tracks.

The Golden orchard tractors, the Dorado, Silver, Explorer II and two models of the SAME Diamond with 230 and 260 hp six-cylinder turbocharged engines were still current in 2004, together with the recently launched Iron and Super Iron. The four 125–163 hp SAME Iron tractors with six-cylinder turbocharged engines with a higher rated horsepower for power take-off and transport work had a forty forward and reverse power shuttle transmission. The Super Iron tractors had 173, 192 and 205 hp engines under their bonnets.

Schlüter

Anton Schlüter was making small and medium-sized diesel engines at Munich for several years before the first Schlüter DZM14 tractor with a single-cylinder four-stroke engine appeared in 1937. Single- and twin-cylinder 15 and 25 hp models with a four forward and one reverse gearbox were added in 1938 and a few gas-powered tractors were made during the war period. Full tractor production resumed in 1949 when single- and twin-cylinder 15 and 25 hp engines were used for the five forward speed Schlüter DS15 and the DS25 with seven forward gears. The three-cylinder 45 hp Schlüter with six forward speeds and one reverse, live power take-off and hydraulics was added in 1959. Engine power moved into the 24–80 hp bracket in the early 1960s when the tractor range included the Schlüter 'Big Four' and three Allrad

280. The late 1960s Schlüter Super 1250 had a 120 hp six-cylinder diesel engine under the bonnet. (Stuart Gibbard)

models. The Schlüter 'Big Four' consisted of the two- and four-wheel drive S650/S650V and the S900/S900V. Mid-1960s sales literature described them as thoroughly modern tractors with high-speed multi-cylinder engines, a twelve-speed column gear change, fully independent power take-off and armchair comfort for the driver. The leaflet continued, 'We saw to it that the de luxe Schlüter farmer's seat was not only the finest example of comfortable tractor seats available but that it also formed an elegant part of all Schlüter tractors – this is what makes them so special.' The fully upholstered seat with parallel spring suspension and a hydraulic damper was also said to help ensure a reduction in spinal problems so that drivers would feel relaxed and comfortable. There were levelling boxes on both hydraulic linkage lift rods and the power steering was so light that the tractor could be 'steered with one finger even on the most difficult terrain'. Sales literature also explained that sophisticated technical extras and modern armchair comfort had been added to the 'Bear Strength' of their tractors and concluded that 'Schlüter builds tractors for those discerning farmers for whom only the best is good enough.'

Reco-Schlutrac, a subsidiary of Rustons Engineering Co Ltd at Huntingdon, chose the 1978 Power in Action event in Suffolk for the first public demonstration of the 160 hp four-wheel drive Schlüter Super E 7800TV with unequal-sized wheels and a £22,060 price tag. A Reco-Schlutrac advertisement in 1979 informed the farming public that 'Tomorrow's tractor is here today.' The Schlüter Compact, Super and Profi-Trac tractors had long-stroke, low-revving engines and the Hydramatic clutch was said to cushion load variations on the engine and give a threefold increase the life of the mechanical clutch.

Several Schlüter tractors including 80 and 95 hp two- and four-wheel drive compact models with Hydromatic fluid clutches and a Super-silent cab were on display at the Royal Smithfield Show in 1980 and 1981. Described as the cost savers the Schlüter range had low-revving, long-stroke engines hand built for longer life, the bonnet hinged forward and the cab could be tilted with a hydraulic ram for easy servicing. Ruston Engineering's catalogue entry for the 1982 Royal Smithfield Show included a range of high-quality Schlüter tractors from 80–240 hp. The 240 hp Super 2500VL was hand built in small production runs. It had a Hydromatic clutch designed to reduce wear in the transmission system

281. The Euro Trac systems tractor was introduced by Schlüter and Claas in 1990.

and a hydraulic ram to tilt the cab. Although Schlüter tractors were available in the UK until 1984 very few were sold to British farmers.

Schlüter introduced a prototype Eurotrac 1600 high-speed systems tractor in a joint venture with Claas at the 1991 Agritechnica Show in Germany. Eight models of the Eurotrac in the 80–190 hp bracket were planned and it was reported in the farming press that fifty tractors were to be available for sale in 1991. The six-cylinder 160 hp engine on the prototype four-wheel drive Eurotrac 1600 was located below and behind the high-level, centrally mounted cab with reversible controls for bi-directional operation and it could be tilted sideways hydraulically for engine maintenance purposes. The Eurotrac had equal-sized wheels and was designed for use with front- and rear-mounted equipment. Provision was made to vary the tractor's weight distribution by moving a 1½ tonne metal ballast weight backwards and forwards with a hydraulic ram

but little more was heard of the Eurotrac design until Claas introduced the similar Xerion systems tractor in 1997.

Singer

Singer Motors bought the manufacturing rights for the four-wheeled OTA Monarch Mk III tractor from Oak Tree Appliances in 1953 when production was moved to Birmingham. The renamed Singer Monarch had a four-cylinder 17 bhp petrol or tvo Ford industrial engine which used about half a gallon in an hour, a six forward and two reverse dual range gearbox, a four-speed power take-off and a belt pulley. A new orange colour scheme was the most obvious change when the improved Monarch Mk IV with a standard category I three-point linkage appeared in 1955. About 1,000 OTA and Singer Monarch tractors had been made when the Rootes Group bought Singer Motors in 1956 and closed the tractor production line.

282. Sales literature described the Singer Monarch as a tractor that 'sets a new record in economy and efficient use'.

Skid Units

Tractors supplied to another manufacturer without their wheels and delivered on a pallet or a skid were generally known as skid units. Some manufacturers used them or a complete tractor to provide power for self-propelled farm machines. The Howard Dungledozer built around a Fordson standard tractor with Rotaped tracks and used to clean out stockyards and load the manure into a trailer was an early example of this idea. The Dungledozer loaded between 12 and 25 tons of manure in an hour and the makers suggested that even higher work rates were possible if at least three tractor trailers or six horse carts were available.

The mid-1950s self-propelled high-clearance Standen Crophopper with a 30–40 hp tractor skid unit was used to hoe, spray and cultivate sugar beet and other rowcrops. A Farmall M provided the power for the mid-1940s American International Harvester HM-1 sugar beet harvester and the self-propelled Shotbolt potato harvester was built around a Nuffield Universal tractor. Claas, Dechentrieter, Ferguson, JF and others wrapped combine harvesters around various makes of tractor. It was claimed that the combine unit could be attached or removed in a few hours.

283. The Standen Crophopper could be used to drill, hoe and spray rowcrops. At the end of the season it was possible to remove the Massey Ferguson skid unit, put it back on its wheels and use the tractor for other work.

284. Catchpole Engineering used a Nuffield tractor with the front wheels removed for the self-propelled Shotbolt potato harvester.

Peter Standen, Catchpole Engineering Co, Johnsons Engineering, John Salmon and Ransomes used tractor skid units in the 1960s and 1970s for their one- or two-row self-propelled root harvesters. The Peter Standen used a Massey Ferguson 35 or 135 for the Solobeet, a Ford 3000 provided the power for the Ransomes Powerbeet and Johnsons Engineering chose a Fordson Major tractor for their self-propelled Johnson Twin Major potato harvester. The farmer customer sometimes provided the tractor for his new harvester and occasionally took the trouble to remove it at the end of the season and use the tractor for other work.

Smallholder

The Smallholder Tractor Co at Evesham, which took over the range of Sutcliff tractors in late 1997, continued the production of the ST20 and the ST35 tractors based on the Ferguson TE20 and FE35. The smaller three-cylinder 30 hp ST20 with a Lister-Petter direct injection diesel engine had a four forward and one reverse constant mesh gearbox. There was a choice of a 30 hp three-cylinder or a 41 hp four-cylinder water-cooled Lister-Petter diesel engine for the Smallholder ST35 with a dual-range six forward and two reverse speed transmission. Smallholder tractors were made with a mix of new and reconditioned parts and prices in 1998 ranged from £5,000 for the ST20 to £6,750 for the four-cylinder ST35.

The ST20 and ST35 were joined by the new four-wheel drive ST28 and two-wheel drive ST135 on the Smallholder Tractor stand at the 1998 Royal Smithfield Show. The 28 hp ST28 had nine forward and three reverse gears, power steering, wet internal brakes, live hydraulics and a live two-speed power take-off. The 41 hp Smallholder ST135 and 55 hp ST135T, based on refurbished MF135 tractors, had direct injection Lister-Petter diesel engines under their fibreglass bonnets, a dual-range transmission with optional creeper box, Ferguson hydraulics and a diff-lock.

285. The Peter Standen Solobeet sugar beet harvester, launched in 1964, could be delivered with a new Massey Ferguson, Ford or International tractor skid unit or with one supplied by the farmer.

The 2001 Smallholder range included three four-wheel drive and two two-wheel drive tractors. The 40 and 48 hp two-wheel drive ST2040 and ST2045 models had three-cylinder Perkins diesel engines. The specifications of the four-wheel drive three-cylinder ST2040, the ST2045 and the 65 hp four-cylinder ST2050 included an eight forward and reverse high/low transmission, a roll bar, a square bonnet and a trailer tipping pipe.

From 2002 the Smallholder Tractor Co used Lombardini direct injection diesel engines for the two- and four-wheel drive 2040 and 2050 tractors. The three-cylinder Smallholder 2040 and the four-cylinder 2050 had 52 hp water-cooled engines, a dual clutch, a twelve forward and reverse synchronised shuttle transmission, hydrostatic steering, dry inboard drum brakes and a fold-down roll bar. A pick-up hitch and a luxury steel and glass weather cab with a radio and air conditioning were optional extras. Sales literature informed potential customers that in order to prove their reliability a Smallholder tractor was driven the 1,007 miles from John O'Groats to Lands End in March 2000 in four-and-a-half days.

Steiger

John Steiger and his sons Douglas and Maurice built a big tractor with a 238 hp Detroit diesel engine on their Minnesota farm in 1957. After using the tractor on their farm for several years the Steiger family decided in 1963 to build five more tractors, which they sold to neighbouring farmers. At least 100 Steiger tractors were being used on American and Canadian farms in 1969 when the Steiger brothers moved to Fargo in North Dakota. By the early 1970s the Steiger range included the Wildcat, Super Wildcat, Cougar, Bearcat and Turbo Tiger. The specification included a General Motors diesel engine, a shift-on-the-move transmission, power steering and a heavy-duty drawbar but hydraulic linkage and power take-off were not included in the basic price of the tractors.

286. There was a choice of a 30 or 41 hp Lister-Petter diesel engine in 1998 for the Smallholder ST 35 tractor.

Steiger also made the four-wheel drive Allis-Chalmers 440 with a 208 hp Cummins V8 engine in the early 1970s and the Ford FW30 introduced to British farmers in 1978 was a 265 hp Steiger with Ford blue paintwork. The 335 hp Steiger was sold in the UK as the Ford FW60 but the FW20 and the FW40 versions of 210 and 295 hp Steigers did not cross the Atlantic.

Turbocharged six-cylinder Caterpillar engines provided the power for the mid-1970s 270 hp Cougar ST270, the 325 hp Panther ST325 and the 225 hp Bearcat PT225. The three tractors also had a hydrostatically driven and electronically controlled power take-off protected by an automatic overload shutdown device.

Offchurch Tractors Ltd at Leamington Spa imported a range of Steiger tractors, including the Cougar ST251, Panther ST310 and the Panther ST350, for the first time in 1976. They had Cummins 250, 310 and 350 hp turbocharged engines, a twenty forward and four reverse gearbox, an air-conditioned cab and optional category III hydraulic linkage. Offchurch Tractors, later trading as Agripower, were still importing Steiger Tractors in the 225–450 hp bracket in 1979. They included the new 450 hp four-wheel drive Tiger ST-450 with a six-speed manual powershift transmission, a cab with full instrumentation mounted on anti-vibration rubber blocks and fourteen work lights.

The Panther 2000, introduced in 1982, was the first Steiger tractor with a twelve-speed full powershift transmission, electronic controls and a de luxe cab. The 525 hp Tiger 'KP' added in 1983 had a twenty-four speed power shift transmission and the Puma 1000 launched in 1986 was the first Steiger with articulated steering and a steerable front axle designed for rowcrop work.

When Steiger Tractor Inc became bankrupt in 1986 Tenneco, the Case IH parent company, bought the business. The first red Case IH Steiger 9100 Series tractors were made in 1988 and the last lime-green Steiger tractors were sold in 1989. The more powerful Case IH 9200 Series appeared in 1990 followed by the 9300 Series in 1995 and the Case IH Steiger Quadtrac with an independently suspended track at each corner was introduced to American farmers in 1996.

287. Offchurch Tractors at Leamington Spa imported the Steiger Cougar in the late 1970s.

The Steiger name was reintroduced to the UK in 1996 when the Case IH Steiger 9300 series articulated tractors were demonstrated to British farmers. The first rubber-tracked Quadtracs were added to the UK Steiger range of Case IH tractors in 1997 and four models of a new generation of articulated Steiger STX Series were launched in 2000.

Steyr

Steyr, a weapons manufacturing business established in the Austrian town of the same name in 1864, added cars, trucks and motorcycles to its product range in the late 1890s. The first Steyr farm tractors were made in 1915 and in the mid-1930s Steyr made engines for Daimler in Germany. Steyr took no further interest in tractor production until becoming part of the Steyr-Daimler-Puch organisation in 1947.

Four-stroke water-cooled diesel truck engines were used for the first postwar Steyr tractors, which had a rather old-fashioned appearance. Regular and high-clearance versions of the 15 hp Steyr 80 Junior had a single-cylinder engine, a four forward and one reverse gearbox and power take-off. A rear-mounted belt pulley and hydraulic linkage were optional extras. The later Steyr 180 with a 30 hp twin-cylinder diesel engine, five forward gears and one reverse was sold in considerable numbers in Austria and continental Europe. The Steyr 180 with a top speed of 15 mph and leaf-spring suspension on the front axle was considered an ideal tractor for road transport work.

More powerful models made at the St Valentin factory in Austria in 1952 included the three-cylinder 45 hp Steyr 185 with six or nine forward and two reverse gears and the four-cylinder 60 hp Steyr 280 with a seven forward and two reverse gearbox. The 185A with a 55 hp engine and 68 hp 280A were added in the mid-1950s.

The Steyr 180, 190, 280 and 290 were made in the early and mid-1960s and special Jubilee models appeared in 1964 to celebrate Steyr's centenary year. Steyr-Daimler-Puch opened a depot at Nottingham in 1967 and the Steyr-Plus series launched that year were the first Steyr tractors sold in the UK. There were six basic models which became fourteen when four-wheel drive and special versions were taken into account. Water-cooled direct injection engines with a heater plug in each cylinder for cold starting and a 'frost-proof maintenance-free radiator with long-life filling' were used across the range. The Steyr-Plus 86ST, 430,

288. The 57 hp Steyr-Plus 650 – one of the first Steyr tractors sold in the UK in 1967 – had live hydraulics and a 540/1,000 rpm power take-off.

540 and 650 had one-, two-, three- and four-cylinder engines rated at 18, 33, 44, and 57 hp respectively. The six-cylinder 77 hp Steyr-Plus 870 and 100 hp Steyr 1090 had a twelve forward and six reverse synchromesh gearbox and a two-speed power take-off. The letter 'a' after the model number denoted four-wheel drive and this applied to some of the more powerful Steyr-Plus tractors including the 870a and 1090a. Sales literature pointed out that the 'Steyr comfort seat leaves absolutely nothing to be desired and can be adjusted without tools in accordance with the weight of the driver' and 'a hydraulic shock absorber with the excellent spring of the seat shields the body of the driver from unpleasant jolts'.

The Steyr Haflinger advertised as a four-wheel drive cross-country vehicle with superb climbing ability and stability on an incline of up to 45 degrees and a 14 in fording depth appeared in the late 1960s. Ryders Autoservice at Bootle marketed the Steyr Haflinger in the UK. The specification included an air-cooled engine, a five-speed gearbox with a top speed of 50 mph, a trailer drawbar with a 10 cwt payload, a canvas or fibreglass cab and an optional power take-off.

The six-model range in 1973 included the more powerful four-cylinder 40 hp Steyr 540, the 60 hp 760/760a and the six-cylinder 98 hp 1100/1100a with twelve forward and six reverse gears and independent power take-off. Steyr-Daimler-Puch joined the big tractor league in 1974 when they launched the two- and four-wheel drive Steyr 1200/1200a and the four-wheel drive Steyr 1400a with turbocharged six-cylinder water-cooled engines and lower link sensing hydraulics. The 155 gross hp 1400a had a twelve forward speed synchromesh gearbox with an optional change-under-load transmission that gave a 32 per cent speed reduction in each gear. An optional creep speed box increased the gear range to thirty-six forward and sixteen reverse ratios. Other features included a hydraulically operated diff-lock on both axles and a 1,000 rpm power take-off shaft rated at 130 hp. The 135 hp Steyr 1200/1200a had the same twelve forward and six reverse transmission as the 1100/1100a, hydraulic brakes on the driven wheels and live power take-off with ground speed.

The 760 with manual steering, shoe brakes and independent hydraulic linkage was the only Steyr

289. The 90 hp Steyr 8110 had a synchronised three-range gearbox with twenty-four forward and six reverse gears.

model listed in the 1976 Green Book. The 66 SAE hp 760 had an eight-speed gearbox with a reversing mechanism to give four speeds in each direction in the field range and four more speeds for roadwork.

W Bridgeman & Son at Newbury in Berkshire imported the red and white Steyr 545, 760, 980, 1100 and 1400a tractors with 45–140 hp water-cooled direct injection engines in 1977 and some were exhibited at that year's Royal Smithfield Show. The improved 760 had the widest range of gears with sixteen forward and eight reverse speeds but the 1400a with a twelve forward and four reverse gearbox had the fewest gears in the Steyr tractor range. Bridgeman & Son were marketing the Steyr 8000 series including the four-wheel drive 8100a and 8140a in 1978 and 1979 but they changed their allegiance to Hürlimann in 1980 which meant that the Austrian tractors were no longer available in the UK.

Steyr introduced the 150 hp 8170A with two assistor rams to boost the rear linkage lift capacity to 6½ tons and the 140 hp 8106A in 1980. The 70–150 hp four-wheel drive Steyr tractors with a sixteen, eighteen or thirty-six forward speed gearbox were imported for a while in the early 1980s by Bexwell Tractors at Downham Market in Norfolk. Introduced by Steyr-

Daimler-Puch in 1984 the 64 hp Steyr 8075 was, like its earlier 48 hp Steyr 8055 stablemate, a lightweight tractor with a low centre of gravity, sixteen forward speeds and an economy cab.

Another chapter in Steyr's history in the UK started in 1989 when Steyr-Daimler-Puch and Marshall Tractors joined forces to establish Marshall-Daimler Ltd at Scunthorpe. The new company introduced the Steyr 8000 range with 64–150 hp engines, which were sold in the UK as the Marshall D series with gold and black paintwork. Introduced in 1990, the 56, 64 and 72 hp Marshall S tractors with low-profile cabs were also made by Steyr. However, with a marked decline in UK tractor sales the Marshall-Steyr partnership came to an end in 1991.

The Austrian tractors with their familiar red and white paintwork reappeared later that year when Morris, Corfield & Co at Much Wenlock in Shropshire were appointed Steyr distributors for the UK. Three 64–80 hp two- and four-wheel drive tractors and four 95–150 hp four-wheel drive models were available in time for the 1991 Royal Smithfield Show. Sales leaflets explained that the comprehensively equipped and technologically advanced tractors lived up to the Steyr motto of 'quality without compromise'. The Steyr 900 series, added in 1992, had environmentally friendly engines that could run on rapeseed oil, a catalytic converter was available at extra cost and biological oil was approved for use in the hydraulic system. The five 42–70 hp Steyr 900 series tractors had sixteen forward and eight reverse gears and a cab low enough to 'fit into the lowest barn'.

The Steyr 9000 series with 78, 86 and 94 hp four-cylinder MWM engines replaced the 8000 series in 1993. There was a choice of mechanical or electronic hydraulic linkage control and with a plastic container incorporated in the hydraulic pipework to collect spillages the driver was able to disconnect the external hose couplings while they were under pressure. Oil level sight tubes for the gearbox and hydraulic system were another feature on the new tractors.

The two- and four-wheel drive Multi-trac range, based on the 9000 series, was introduced in 1994. The M968, M975 and M9083 had 68, 75 and 83 hp turbocharged engines respectively under their steep sloping bonnets. A shuttle transmission or optional powershift gearbox together with an integrated front power take-off and hydraulic linkage were features of

the Steyr Multi-trac. The 94 hp M9094 was added later in the year. The 320 hp 9320a Power Trac, also announced in 1994, was the flagship of the Steyr range. It had a hydrostatic transmission combined with a conventional four-speed gearbox and a reversible driving position, making the four-wheel drive tractor equally efficient in both directions.

The Steyr 9100 range, launched in 1996, included the 9105a, 9115a, 9125a and 9145a with the last three numbers denoting horsepower. These four-wheel drive tractors had push-button controls for the twenty-four forward and reverse speed powershift transmission, hydraulic linkage and hydraulic spool valves. An optional power take-off management system could be used to stop or start the power shaft automatically when raising or lowering the hydraulic linkage.

When the Case Corporation acquired Steyr Landmaschinentechnik in 1996 the Austrian-built tractors were still imported for a while by Morris Corfield who sold them through their existing UK dealer network. The 94 hp, sixteen-speed Case CS94 and the 150 hp CS150 with a forty-speed forward and reverse transmission and a computerised headland

290. *The 115 hp Steyr M9145, exhibited at the 2000 Royal Smithfield Show, had computerised headland and power take-off management systems.*

management system launched in 1996 were the first Steyr tractors sold in Case IH livery.

Bonhill Engineering Ltd, at Beverley near Hull, became Steyr distributors in 1998. Previously at Brough, Bonhill imported Fendt tractors until 1997 when AGCO bought the Fendt tractor business.

Steyr M900, M9000, M9100 and CVT models were included in the Case New Holland product range for 2001. Four Steyr M900 models with 48–75 hp three-cylinder turbocharged models had a sixteen forward and eight reverse speed synchro shuttle transmission. The three 78, 86 and 94 hp M9000 series four-wheel drive tractors had a sixteen forward and reverse speed transmission. The four-wheel drive 120–170 hp Steyr CVT series tractors with constantly variable transmission had six-cylinder turbocharged engines and the 110–150 hp 9100 tractors were equipped with thirty-two forward and reverse transmissions.

Stockhold

HJ Stockton Ltd of London decided that there was still demand in the late 1950s for a relatively basic lightweight tractor similar to the earlier BMB President which had a Morris vaporising oil engine, a three forward and one reverse gearbox and optional hydraulic linkage. The Stockhold President was launched at the 1957 Royal Smithfield Show and with the exception of its twin-cylinder air-cooled 14 hp Petter diesel engine, the new tractor was similar to the BMB President which had been discontinued in 1956. The Stockhold President cost about £360 but electric starting and lights, power take-off and hydraulic linkage were optional extras. Demand did not come up to expectations and very few were sold.

Sutcliff

The Sutcliff ST 230C Mini-Trak A, made partly from recycled Ferguson TE20 components with a moulded plastic bonnet, wings and transmission cover, was introduced by Sutcliff Industries at Evesham in Worcestershire in 1995. The mini-tractor, which cost £5,800, had a three-cylinder 30 hp or four-cylinder 41 hp Lister-Petter direct injection diesel engine and a single-plate dry clutch. The four forward and one reverse gearbox with optional creep speed and disc brakes were installed in a reconditioned Ferguson TE20 transmission housing. Optional extras included an independent power take-off, a lighting kit and a safety frame or weather cab.

291. The Stockhold President, similar to the earlier BMB President, was introduced in 1957 but very few of them were sold.

292. A safety roll bar, weather cab, downswept exhaust and a lighting kit were optional extras for the Sutcliff Mini-Trak made by Sutcliff Industries at Evesham.

The ST35, added to the Sutcliff range in 1997, had a six forward and two reverse Massey Ferguson 35 transmission and a 30 or 41 hp Lister-Petter water-cooled diesel engine. The ST35 was also a mixture of re-manufactured and new parts, including a spring suspension seat and new body panels. The Smallholder Tractor Co, also at Evesham, acquired Sutcliff Industries in 1997.

TAFE

Tractors UK at Sturminster Newton in Dorset was originally established in collaboration with Massey Ferguson in 1961 to market a complete range of TAFE tractors, implements and accessories on the export market. TAFE are the initials of Tractors and Farm Equipment Ltd and their products reflected the design of the earlier top selling MF35 and 135 tractors.

Sales literature in the late 1990s explained that TAFE tractors, built with all new components, were reliable, economical and well suited to meet the needs of smallholders and market gardeners. TAFE made in excess of 35,000 tractors for worldwide sale in 2002 when a four-wheel drive and five two-wheel drive models were available in the UK. Category I and II three-point linkage, power take-off and a trailer tipping pipe were standard; optional extras included a pick-up hitch, a folding roll bar and a cab. The 30DI Classic, 35DI Classic and 35DI had a dual clutch, an eight forward and two reverse gearbox, adjustable wheel track settings and a spring suspension seat. The 30 hp 30DI and the 37 hp 35DI had three-cylinder direct injection Perkins engines under their rounded red FE35 bonnets. The 35DI with a dry air cleaner and hydrostatic power steering cost £7,450 but the cheaper 30DI and 30DI Classic had mechanical steering and an oil bath air cleaner. A 47 hp Perkins engine was used for the TAFE 45DI, 45DI OIB and four-wheel drive 45DI OIB with a

293. The TAFE 45DI (left) and 35DI had hydrostatic steering, an eight forward and two reverse gearbox, draft and position control hydraulics and a live power take-off.

square bonnet and rectangular radiator grille. The specification was very similar to the 37 hp tractors but the 45DI OIB had oil-immersed disc brakes.

Later TAFE 65DI, 75DI and 95TDI tractors with the horsepower of the Tier 2 Perkins engine matching the model numbers had a dual clutch, eight forward and two reverse gears, oil-immersed disc brakes, hydrostatic power steering and category II hydraulic linkage. The standard specification for the TAFE 90TDI included a 90 hp turbocharged engine, larger rear tyres and front weights. The 2004 TAFE price list included the four-wheel drive 45DI at £10,950 and the two-wheel drive model was £9,150.

Tractor Vaporising Oil

Tractor Vaporising Oil was a Shell trademark but tvo soon became the adopted name for vaporising oil. Although cheaper than petrol it was less efficient and a vaporiser was needed to preheat the fuel before it was drawn into the engine. White smoke from the exhaust pipe was a sure sign that the engine was too cold to run on vaporising oil and unburned fuel was probably finding its way into the oil in the sump. Wartime restrictions led to a severe shortage of petrol for starting farm tractors so drivers would cover the radiator with an old coat or a corn sack during the winter months to keep the engine warm enough to run on tvo when they stopped for a break.

Esso Green, Regent Gold and Shellspark were the more popular brands of tractor vaporising oil. The Shell and BP Shellspark, coloured purple for easy identification, was claimed to have rapid warm-up properties, improved throttle response and increased tractive effort without engine knock.

It was usual for a vaporising oil engine to have a preheating system, which made use of the heat from the exhaust to warm the fuel before it reached the cylinders. Some tractors had an adjustable heat valve in the exhaust manifold, which released the exhaust gases direct to the exhaust pipe on hot days and circulated them around the inlet manifold when it was cold.

Several companies, including the Beccles Engineering Company, Fishleigh Ltd and the

Loddon Engineering Company, made vaporising oil conversion kits for the Ferguson TEA20 and other petrol-engined tractors. The Fishleigh vaporiser had a modified carburettor, a small petrol tank and a replacement manifold which pre-heated the mixture of fuel and air before it reached the cylinders. The instruction leaflet reminded the driver to make sure the engine was hot enough before switching to vaporising oil.

Trantor

A survey of farm tractor use in the early 1970s found that as much as 70 per cent of the work done by so-called 'ploughing tractors' was either of a general-purpose nature or hauling produce and materials around the farm. This prompted Graham Edwards and Stuart Taylor to design and build a prototype two-wheel drive fast haulage tractor in 1972. The Trantor Transport tractor with a 55 hp Perkins engine was similar in shape and size to a Land Rover but with category I hydraulic three-point linkage and a power take-off. The Trantor had Land Rover wheels at the front, Unimog wheels at the rear and a passenger seat on each side of the central driving seat.

Several prototypes were thoroughly tested and a modified version of the Trantor was exhibited at the 1976 Royal Show. More trials were carried out and the production model of the two-wheel drive Series I Trantor was launched at the 1978 Royal Smithfield Show. The first batch of Trantors, which cost £9,750 complete with a safety cab and suspension on all four wheels, were made by Eva Industries Ltd at Manchester. The Trantor had an 80 hp Perkins 4.236 diesel engine, a synchromesh gearbox and a transfer box to provide ten forward and two reverse gears and a top speed of 60 mph. Four-wheel air-over-hydraulic brakes, an air braking system for a trailer and an independently suspended pick-up hitch made the Trantor an ideal tractor for high-speed road transport. A diff-lock, hydraulic linkage and a two-speed 60 hp power take-off shaft meant that apart from deep ploughing the Trantor was suitable for most field operations. It was classed as a farm tractor with the added advantage of qualifying for an agricultural road fund licence that only cost £8.50 at the time

Shellspark

the best v.o. you can get!

Shellspark is the new, improved tractor vaporising oil. It has quicker warming-up properties that make it second-to-none. And its advantages include both *economy* and *performance* :—

Shellspark saves starting petrol — because it has quicker warming-up properties.

Shellspark improves flexibility and throttle response.

Shellspark gives better combustion, keeps plugs cleaner and eliminates misfiring.

Shellspark gives increased tractive effort without engine knock.

Shellspark sustains maximum power for belt work, with less wear and tear.

Shellspark is coloured purple for easy identification and to prevent misrepresentation.

CHANGE OVER TO SHELLSPARK

Order this fine new fuel now, from your local Authorised Distributors for

SHELL - MEX AND B.P. LTD.

SHELL AND BP FARM SERVICE BP

294. This advertisement for Shellspark vaporising oil appeared in 1955.

and it was also legal to run the Trantor on tax-rebated farm fuel.

The two-wheel drive Series II Mark I Trantor launched in 1983 had a turbocharged Leyland engine. Wheel track width was adjustable in 4 in steps from 64–80 in and the cab was repositioned to make space on the rear load platform for a de-mountable crop sprayer or fertiliser spreader. Depending on the type of tyres used the Series II was, according a sales leaflet 'a low ground pressure tractor suitable for all year round spraying and spreading at 12 mph or as a high tractor for road haulage at 40 mph'. Production of the 752, 952 and 1252 Trantors rated at 54½, 71 and 92 hp with a ten- or twenty-speed gearbox and a 12,000 mile or twelve-month warranty was planned to start in the mid-1980s but none were built.

295. The Series I Trantor fast tractor had independent front axle suspension and leaf spring suspension on the rear axle.

296. The Series II Trantor with a Leyland diesel engine had a top road speed of 40 mph.

297. The four-wheel drive Trantor Hauler had a six-cylinder 126 hp Perkins diesel engine.

The Trantor Hauler with a six-cylinder 126 hp Perkins engine launched at the 1984 Royal Show had an on-line trailer braking system to meet new Department of Transport regulations. It had a 96 hp turbocharged Leyland engine but a special model with an 80 hp Perkins engine was made for local authority use. A bogie unit enabling the Trantor to pull articulated lorry trailers was added in 1991.

A new Euro-Trantor was on the drawing board when the company changed its name to HST (High Speed Tractors) Developments Ltd in 1988 and the first prototype Euro-Trantor was made in 1996.

Although the Trantor was not particularly popular with British farmers several hundred had been sold when production was transferred to India in the mid-1990s where it was made by Eicher. India and China were seen as countries with a growing demand for high-speed haulage tractors and annual tractor sales were in excess of 200,000 units in the late 1990s. A new four-wheel drive Trantor 904 also

made by Eicher, with an Indian-built 90 hp Tata engine, was introduced in 1998.

Trusty

Tractors (London) Ltd of Bentley Heath in Hertfordshire, which made the first pedestrian-controlled Trusty garden tractors in the mid-1930s, introduced the four-wheel Trusty Steed in 1948. The first Steed, which cost £200, was little more than a ride-on towing tractor with an 8.9 hp Douglas engine, a centrifugal clutch, and a two forward and one reverse gearbox mounted behind the driver's seat. Sales literature explained that the Steed, with a top speed of 8 mph, had 'a centrifugal clutch to ensure a silky start at all times'. A separate drive to each rear wheel gave a very short turning radius, which made it possible for the 'handy and easily controlled little machine to pull a man-sized load and be readily manoeuvrable in a restricted space'.

A more sophisticated Trusty Steed introduced in

298. Introduced in 1950 the 14½ hp Trusty Steed with a JAP or Norton engine had three forward gears and one in reverse.

1950 with adjustable wheel track and a mid- or rear-mounted toolbar raised and lowered with a spring-assisted hand lever cost £205. The new Steed had a front-mounted 14½ hp JAP or a Norton air-cooled engine, a single-plate clutch, three forward gears and one in reverse, independent brakes and power take-off. An engine bonnet was available and there was a choice of a vertical or horizontal exhaust pipe. A few Trusty Steeds on Roadless half-tracks were sold for work on marginal or hilly ground. About 500 Steeds were built in the 1950s and 1960s and a few more, with occasional improvements, were produced before the last one was made in 1978.

The Trusty Tractor Co, still at Bentley Heath, re-introduced the Trusty Steed in 1995. The new four-wheel drive Steed, made in the Czech Republic, had a 33½ hp Lombardini three-cylinder water-cooled diesel engine and a hydraulically

299. The four-wheel drive Trusty Steed was made in the Czech Republic in 1995.

operated single-plate clutch. Other features included hydraulic disc brakes, hydrostatic steering, live front and rear category I hydraulic linkage with a separate oil reservoir and a de luxe safety cab with optional air conditioning.

Turner

The Turner Manufacturing Co at the Wulfruna Works in Wolverhampton had been involved in the motor vehicle industry since the early 1900s when they launched the green and yellow Turner diesel tractor at the Royal Show in Shrewsbury in 1949. The tractor became known as the Turner Yeoman of England but this name was not used for early sales leaflets. The Yeoman of England, which cost £690, was described as a '40 hp diesel-engined wheeled tractor designed to be used with trailed or mounted implements'. It had a Turner four-cylinder V-form engine, a single-plate dry clutch, four forward gears and one reverse, spiral bevel

differential, spur gear final drive, independent foot brakes, hydraulic linkage and the wheel track was adjustable in 2 in steps from 48–72 in. Power take-off, belt pulley and lights were optional equipment.

Manufacturers' sales literature in the 1940s and 1950s was usually generous in the praise of their products and Turners were obviously proud of their tractor. It was claimed that 'the Yeoman of England would do much more work at less cost per acre and its tremendous lugging power, traction and economy in operation are some of the many advantages enjoyed by farmers using a Turner diesel tractor'. Potential purchasers were also told that 'the astonishing lugging power of the Turner diesel engine enables the tractor to pull four furrows as easily as other tractors are able to pull three – thus giving the farmer one extra ploughed acre in an hour without additional fuel costs'.

Although an electric starter and a decompressor were included in the basic price the tractor was

300. The offset seat and steering wheel on the 40 hp Turner Yeoman of England gave the driver a clear view when working in rowcrops.

sometimes difficult to start in cold weather and because of this and other reliability problems the Yeoman of England did not prove popular and the last tractor was made in 1955.

TYM

Motokov UK added the distributorship of the South Korean TYM tractor range to their existing Zetor, Landini and EuroTrac franchises in 2000. The 39 and 55 hp TYM tractors with a flat floor driving platform and safety roll bar were launched at the 2000 Royal Highland Show. The smaller TYM T390 with a four-cylinder Kubota engine had a twelve forward and reverse synchroshuttle transmission and the T550 with a three-cylinder 55 hp John Deere engine was equipped with a synchroshuttle sixteen forward and reverse gearbox with creep speeds. Both tractors had position and draft control three-point linkage, power steering and an independent two-speed power take-off with an electric safety switch which disengaged the

power take-off when the driver raised the hydraulic linkage. The T550 with a roll bar cost £13,685 in 2000 and the optional air-conditioned cab added £1,310 to the price.

The TYM T550 with the choice of a roll bar or safety cab and the TYM T700 made by Tong Yang Moolsan in South Korea were included in the 2004 Motokov UK price list. Powered by a three-cylinder turbocharged 70 hp John Deere engine, the four-wheel drive T700 had a four-range sixteen forward and reverse synchronised shuttle gearbox and an engine-mounted hydraulic pump. The tractor had a computerised headland management system in its spacious air-conditioned cab with a chiller and a heated box for food and drinks. The TYM K Series introduced in 2005 with 54½, 58½ and 65 hp Caterpillar engines had a twenty-four forward and reverse transmission with creep speeds and a computerised headland management system in the air-conditioned cabs.

301. The TYM T390 was made in South Korea and imported by Motokov UK.

Chapter 9

Unimog – Zetor

Unimog

A prototype four-wheel drive fast tractor with equal-sized wheels was built by a group of German engineers and farmers in the mid-1940s. Designed in the style of a military vehicle it had a 25 hp Mercedes-Benz diesel engine, a wide range of forward and reverse speeds, a diff-lock, front and rear power take-off shafts, hydraulic linkage, a rear load-carrying platform and a steel-framed canvas two-seater cab. After completing its field trials in 1946 the first Unimogs – the name derived from Universal Motorised Gerat (vehicle) – were made at Goppingen in Württemberg in 1949.

The Unimog was used in agriculture and forestry and for municipal applications and more than 600 had been made by 1951 when Daimler-Benz bought the design. Production was transferred to Gaggenau in the same year and the 25 hp Unimog 2010 was awarded a silver medal by the German Agricultural Society. Sales literature explained that various implements including toolbar equipment, a front-mounted mower, a hay tedder, a baler, a crop loader and even a binder could be used with the Unimog. It was also suggested that a pto-driven front-mounted binder could be 'operated by

302. Sales literature explained that this 25 hp Unimog with a side-mounted forage harvester and towing a trailer made hard work easy.

303. Launched in 1974 this Unimog had a top road speed of 50 mph.

the driver who will be able to watch and control the apparatus from his seat and no assistant will be necessary'.

The original steel and canvas cab was replaced with an all-steel version when the U600 was launched in 1953 and the 82 hp Unimog S specifically designed for off-road operation appeared in 1955. Daimler-Benz claimed the long wheelbase version of the model S Unimog introduced in 1958 was the first farm tractor with a synchromesh gearbox as standard equipment.

Production passed the 50,000 mark in 1961 and the Unimog 406 with a 65 hp petrol engine was added to the range in 1963. An optional Mercedes-Benz direct injection diesel engine for the 406 was introduced in 1964. The Unimog S was still in production when the Unimog U series, including the U34, U40, U54, U70 and U80, was launched in 1966. The U45, U66 and U90 were added in 1968 and the first 100 hp Unimog appeared in 1969.

Mercedes-Benz introduced the more sophisticated MB-trac in 1971 when an improved and rationalised range of Unimogs included the U52, U66 and U84. Two new 82 hp and 110 hp chassis-type Unimogs with a load platform for mounting a sprayer or fertiliser spreader were launched in the same year.

The U120/425 with a 120 hp engine and an eight-speed gearbox with a top road speed of 50 mph launched in 1974 was the most powerful Unimog yet made by Mercedes-Benz. Optional gearboxes with eight field speeds and/or eight creep speeds giving a total of sixteen or twenty-four forward gears were also made. Other features of the U120 included pneumatically operated diff-locks on both axles and two-speed front and rear power take-off shafts with independent hydraulic clutches. The Unimog reached the 150 hp mark in 1975 with the launch of U1500. The 100 hp U1000 was added in 1976.

Long wheelbase Unimogs including the U1300L and the 168 hp flagship U1700L were added to the fleet in the late 1970s. By 1980 there were eight models on the UK market including four specialist agricultural tractors which shared the same basic engine and some other components with the MB-trac.

Unimogs sold by Mercedes-Benz at Milton Keynes in the mid-1980s ranged from the U600 with a 52 hp engine and twelve forward gears to the 150 hp U1500AG with a standard sixteen forward and reverse gearbox. The short wheelbase U1400 and U1600 with 136 hp and 156 hp diesel engines were added in 1988. Although seven models were in production at Gaggenau in 1990, only the 102 hp U1000, 125 hp U1200 and 170 hp U1700 Unimogs with six-cylinder engines and a sixteen forward and reverse gearbox with a top speed of 47 mph were available in the UK.

304. *The Unimog U1200 with a 125 hp six-cylinder engine was one of three models of Unimog introduced to British farmers in 1990.*

Accurate control of front- and rear-mounted implements on tractors with a full suspension system had been difficult until 1990 when the Unimog EHR hydro-tonic system, with ultra-sound sensors which constantly monitored the working depth or height, solved the problem. The optional Unicomp 'in-cab' computerised engine and performance-monitoring system was also introduced in 1990 and, on a practical note, a much larger rear window in the cab improved the visibility of rear-mounted implements.

The Terramatic electronic linkage control announced in 1994 improved the Unimog's performance with a plough. The Terramatic system monitored variations in oil pressure in the front or rear hydraulic lift cylinders and the system used this information to maintain constant draft and keep the depth as even as possible. It was also used to monitor the load on the rear platform and keep the spray bar, or spreading mechanism, at the correct height as the fertiliser hopper or sprayer tank discharged its contents. The Terramatic could be used

305. *The Terramatic hydraulic draft control system was standard on the mid-1990s Unimog 1200.*

with either the front or rear hydraulic linkage but not with both at the same time. An optional radar-sensing unit was available to control the working depth of trailed machines such as sugar beet and potato harvesters.

Four agricultural Unimogs were available in the UK in 1994. The U90AG, U1400AG and U2100AG, launched at that year's Royal Show, were added to the U1600AG which had been introduced to British farmers in 1991. The 87 hp U90AG with a five-cylinder water-cooled Mercedes-Benz engine had twenty-two forward and eleven reverse gears while the 136, 155 and 214 hp Unimogs with six-cylinder

306. The 115 hp Unimog U90 Turbo with a Mercedes-Benz engine was made between 1998 and 2002.

turbocharged water-cooled engines had a sixteen forward and reverse transmission. Other features included the usual front and rear power take-off and live hydraulic linkages, an engine-driven compressor and selectable two- and four-wheel drive. Right-hand drive was optional for the two-seater cab mounted on rubber dampers.

The smaller Unimog UX100 with a four-cylinder 73 hp or a five-cylinder 120 hp engine and hydrostatic transmission intended mainly for local authority use appeared in 1997. The driving controls could be moved to either side of the cab to suit the work in hand.

The U Trac 160 launched in 1997 and developed by South Cave Tractors at York was a new type of agricultural Unimog. Based on the Unimog U1650 with a centrally mounted cab it had larger tyres to give increased ground clearance for spraying and spreading fertiliser in row crops. The U Trac 160 specification included a six-cylinder Mercedes-Benz engine, a sixteen forward and reverse gearbox and the option of two- or four-wheel drive.

The 1998 Mercedes-Benz Unimog range included four standard agricultural models and two with rear load platforms for a de-mountable sprayer or fertiliser spreader. The 122–214 hp U90AG, U1400AG, U1600AG and U2100AG had Mercedes-Benz engines under their bonnets. Disc brakes and diff-locks on both axles and a 540/1,000 rpm power take-off were standard across the range. The 115 hp U1650 and 214 hp U2150 Unimogs with rear load platforms had eight forward and reverse gears, disc brakes, diff-locks and a power take-off but the hydraulic linkage was an optional extra. The U300, U400 and U500 with more rounded cabs replaced the earlier Unimog range in 2002.

Universal

Universal tractors, introduced to British farmers in 1970, were made by Uzina Tractorul Braşov (UTB). Developed by Fiat in Italy the first Romanian-built four-wheel drive Universal 651M tractors were imported by Universal Tractors at Mitcham in Surrey. The tractor had a 73 hp diesel engine, a ten forward and two reverse gearbox, power-assisted steering, Bosch hydraulics, an independent power take-off and an alternator charging system. The 651M cost £1,420 ex Felixstowe docks and an advertisement explained that it was 'a lot of tractor at any price and £600 cheaper than any other comparable tractor on the market'.

307. Universal Tractors at Mitcham in Surrey introduced the 75 hp Universal 651M to British farmers in 1970.

Iandale Tractors at South Cave on Humberside imported the Romanian-built tractors from 1974 when the range included the Universal U445/U445DT and U550/U550DT two- and four-wheel drive models and the SMU445 crawler. The 45 hp U445 with a three-cylinder Universal engine had a nine forward and three reverse speed gearbox. Features of the 55 hp U550 included a four-cylinder direct injection water-cooled engine, dual clutch, eight forward and two reverse gears with synchromesh on third/fourth and seventh/eighth and power-assisted steering. Both tractors had an independent power take-off, live hydraulics, adjustable wheel track and a 'comprehensive toolkit'. Optional extras included a Duncan safety cab, a belt pulley, rear wheel weights and an alternative final drive bevel gear and pinion giving a 16.7 per cent speed increase in all gears. The two- and four-wheel drive 50 hp Universal 530 and 62 hp 640 were added in 1979. Power steering was standard on the 640 but optional for the Universal 530. The four-wheel drive DT models had a conventional transfer box and prop-shaft drive to the front axle.

308. Iandale Universal Tractors at South Cave on Humberside were importing the 45 and 55 hp U445 and U550 Universal tractors from Romania in the mid-1970s.

Iandale Universal Tractors were still marketing the 45–62 hp two- and four-wheel drive Universal U445, U550 and U640 tractors in 1980 when the price of the U445 was about £4,000. Hydrostatic steering was standard on the four-wheel drive 62 hp U640DT and the two-wheel drive version had power-assisted steering. The SM445 and SM640 crawlers, based on the U445 and U640 with a power take-off and hydraulic linkage, were also available at the time.

The U445, U550 and U640 wheeled tractors and two crawlers with 45 and 69

309. The 70 hp Universal 704DT, marketed by Bonhill Engineering in the mid-1980s, had a four-cylinder direct injection diesel engine.

hp engines were being sold by Universal Tractors at South Cave when the three-cylinder 50 hp Universal U530 was launched in 1981. The Romanian-built tractors were advertised as 'sparkling, highly efficient workhorses with prices from around £4,250'.

New SD models with improved specifications and more driver comfort, described as 'a new breed of workhorses from the Universal stable' were added to the standard two- and four-wheel drive models in 1982. Features of the 445, 530 and 640SD included a semi-flat deck cab, hydrostatic steering, a new Inversor gearbox which had a shuttle reverser with eight speeds in both directions and an assistor ram which increased the lift capacity of the three-point linkage. The importers offered a 2 per cent interest finance plan over twelve months to promote the sale of Universal tractors.

Linx Agriculture at South Cave were selling Lamborghini tractors when they were appointed the Universal tractor distributor for the UK in 1983. The U50DT and U1010DT were launched in 1984 and the U532/U532DT and U642/U642DT replaced the U520

and U640 in 1985. The 101 hp U1010DT with hydrostatic steering and a two-speed power take-off was the first 100 hp Universal tractor on the British market.

Linx Agriculture ceased trading in 1986 and Bonhill Ltd, who were importing German grain and root harvesting machinery at their depot at Thetford in Norfolk, became the UK distributor for Universal tractors. Bonhill retained the Thetford premises for the harvest machinery but the Universal tractor business with a range of 50–100 hp two- and four-wheel drive tractors including high clear and narrow models and two crawlers remained at South Cave.

Four new Universal 3 series tractors with a more streamlined appearance to bring them closer to western European standards were launched in 1988. The two-wheel drive 453, 533, 643 and four-wheel drive (DT) models rated at 45, 53 and 64 hp had twelve forward and three reverse gears with an optional eight-speed shuttle transmission. The 72 hp 723/723 DT had a sixteen forward and four reverse synchromesh gearbox and inboard wet disc brakes. Other features of the Universal

3 series included a much-improved cab with electric screen wipers, a 540/1,000 rpm and ground speed power take-off and hydrostatic steering. A radio cassette player and hydraulic trailer braking system were standard on the 643 and 723. The turbocharged four-cylinder Universal U853 priced at £13,350 was the cheapest 85 hp tractor in Britain when it was introduced in 1990.

310. The mid-1990s Universal 833DT with a full set of road and working lights had a complete set of tools in the tractor toolbox.

Bonhill Engineering were selling the Universal 3 series and a range of Fendt tractors in 1993 but within twelve months Universal Tractors Ltd in North Yorkshire were importing the Romanian tractors. An improved range of three 72–105 hp Universal Euroseries 33 models with new flat floor tiltable cabs, fibreglass bonnets and a four-speed power take-off appeared in 1997. The top-of-the-range four-wheel drive 105 hp six-cylinder Universal 1033 DT had a sixteen forward and four reverse gearbox and the same gearbox was used for the 72 hp 733 DT and the 89 hp 833 DT. A mix of new and improved 45–73 hp Universal Euroseries 3 tractors was also launched in 1997. The three-cylinder 45 hp 453 and 53 hp 533 had an eight forward and reverse shuttle gearbox and the four-wheel drive models had twelve forward and three reverse gears. The same gearbox was used for the 64 hp 643/643DT and the 73 hp 703 DT had a sixteen forward and four reverse gearbox.

Universal Tractor Imports at Ripon in North Yorkshire took over the distribution of the Romanian-built tractors when Universal Tractors Ltd ceased trading in 1997. The 1998 range, still based on former Fiat designs, included seven two- and four-wheel drive models from the three-cylinder 45 hp 453 to the six-cylinder 105 hp Universal 1033.

Ursus

The Polish town of Ursus near Warsaw has been connected with the farm machinery industry since the Zaklady Mechaniczne Ursus (Ursus Mechanical Works) was established there in 1893. Internal combustion engines were being manufactured at the renamed Joint-Stock Society of Special Fittings and Engines factory in 1903. Two- and four-stroke semi-diesel engines up to 60 hp were made at Warsaw in 1912 and before long the range had been extended to include 75–450 hp Ursus diesel engines for power stations and other industrial customers. Following a name change in 1918 to the Ursus Engine and Tractor Factory the first Ursus tractors with a 25 hp twin-cylinder paraffin engine, introduced in 1922, were based on the International Titan.

There was another change of name in 1923, this time to the Ursus Mechanical Works, but the business failed in 1930 and the Ursus Works was nationalised by the Polish government. Cars, buses, trucks, engines and military equipment were built for the government until the late 1930s when tanks, armoured vehicles and Junkers aeroplanes were made there for the German army.

311. Introduced in 1947 the 45 hp Ursus C45 had a hot bulb semi-diesel engine.

The Ursus Mechanical Works was the only tractor factory in Poland when the first 45 hp Ursus LB-45 hot-bulb semi-diesel tractors were made there in 1947. The LB-45, which later became the C-45, could easily be mistaken for a Lanz Bulldog. It had a leaf spring under the front axle and a cushioned driving seat with coil spring suspension, which was offset for better visibility, made the driver's life a bit more comfortable. An electric starter motor was an option for farmers who did not want to spend time warming up the hot bulb with a blowlamp before using the steering wheel to start the engine. The starter motor was used with a trembler coil which provided the necessary sparks to start the engine on petrol and

312. The 50 hp Ursus C350 had ten forward and two reverse gears.

313. An agreement between Ursus and Zetor in the early 1970s resulted in components being shared for some of the tractors made in Czechoslovakia and Poland. The Ursus C-385 and the Zetor Crystal were almost identical with the front end made by Zetor and the back end by Ursus in Poland.

when hot enough it was switched over to paraffin or diesel fuel.

The 25 hp C-325 launched in 1958 was the first postwar farm tractor designed and built by the Ursus Mechanical Works and by the early 1960s the 35 hp Agripol C-335 and the 50 hp C-350 had been added to the range. Ursus became involved with Zetor in Czechoslovakia in a two-way trade of components in 1962 and the Ursus C-4011, launched in 1965, was a mirror image of the earlier 45 hp Zetor 4011. The 80 hp 8011 Zetor Crystal and the Ursus 385 introduced in 1972 with Czechoslovakian engines and Polish transmissions were other examples of co-operation between the two tractor makers.

Maulden Engineering Co Ltd at Flitwick in Bedfordshire introduced the Ursus C-335 and C-350 to British farmers in the mid-1960s. The twin-cylinder 35 hp C-335 with a six forward and two reverse gearbox, live power take-off and Agrocontrol hydraulic linkage and the 50 hp C-350 with a four-cylinder engine had a ten forward and two reverse gearbox, live category II

hydraulic linkage, a diff-lock and electric starting. The four-cylinder 60 hp C-355 had the same gearbox and hydraulic linkage as the C-350 which it replaced in 1970.

Lower link sensing hydraulics were an advanced feature of the two- and four-wheel drive 85 hp C-385 and C-385A launched in 1972. Both tractors had a sixteen forward and eight reverse speed gearbox, power steering, a two-speed power take-off and an air-conditioned cab. The C-385, C-355 and C-335 were still current in 1973 when Ursus tractor production passed the 400,000 mark. At the time the C-385 was priced at £2,995, the C-355 was £1,690 and the C-335 advertised as 'the best value-for-money tractor on the market today' cost £1,358.

Ursus had an annual production of 70,000 tractors for sale on the world market when the two-wheel drive Ursus C-1201 and the four-wheel drive C-1204 with six-cylinder 120 hp water-cooled direct injection engines made their debut at the 1974 Royal Smithfield Show. The C-1204 was the first four-wheel drive Ursus model seen in the UK and both tractors had eight forward and four reverse gears and a torque amplifier which doubled the number of gears in both directions. The torque amplifier used on the move to increase or decrease the forward or reverse speed by about 30 per cent had the opposite effect on drawbar pull. The C-1201 and C-1204 had Plowselect hydraulics with an external assistor ram and a sales leaflet described the cab as 'a luxury model with noise insulation, air conditioning and a passenger seat'. The C-1204 had a spring suspension system on the front axle and a hydraulic clutch that engaged and disengaged front-wheel drive while on the move.

Ursus-Bizon (GB) Ltd of Needham Market in Suffolk imported a four-model range from 1976 which included the Ursus C-1201/1204 and C-385/385A Unified series tractors. They were similar in design with many common parts and a built-in air compressor. The Ursus range marketed by Ursus-Bizon in 1978 was reduced to the 85 and 120 hp models including de luxe two-wheel drive versions with a spring suspension front axle. The four-wheel drive C-1204 cost just over £12,600 and to tempt UK farmers to buy the tractor it was advertised as 'a big-value tractor available through an interest-free finance plan'.

The C-385 and C-1201 were still in production when the new 52 hp two-wheel drive C-362 which cost £5,620 and a de luxe version of the C-385 with independent front suspension were added to the 1980 Ursus price list. The Ursus range was widened again in 1982 with the launch of the 100 hp C-1002/1004 and the six-cylinder 150 hp four-wheel drive C-1604. The C-362 was the only tractor in the Ursus range without a two-speed power take-off, torque amplifier, power steering or a heated cab. Farmers had the choice of a standard or SX model of the 80–150 hp Ursus tractor range in 1982 when additional features on the SX tractors included a de luxe seat, fully adjustable rear wheels, an extra plough lamp and a long/medium wave push-button radio.

Five new Ursus tractors in the 60–160 hp bracket appeared in 1983. The two-wheel drive 60 hp model and the 80, 100 and 120 hp two- and four-wheel drive tractors had a sixteen forward and eight reverse gearbox and the four-wheel drive Ursus 1614 had twelve forward and six reverse speeds. Following a period in the hands of the receiver, Debtrac Ltd, also at Needham Market, bought Ursus-Bizon in 1983 and became the UK distributors for Ursus tractors and other Polish farm equipment. The new owners also acquired the rights to install the Power-Code variable horsepower option, developed in the early 1980s by the Ursus distributor in Holland. Power-Code, similar to the Caterpillar VHP system which was optional for the seven 80–150 hp Ursus tractors available in Britain at the time, allowed the driver to select a lower level of engine power when the tractor was used for light work.

A new factory was built near the Ursus plant in the

314. The 120 hp Ursus 1201 and four-wheel drive 1204 were launched at the 1974 Royal Smithfield Show.

mid-1980s to manufacture Perkins engines and Massey Ferguson tractors under licence. Distribution of Ursus tractors in Britain passed into the hands of three farm machinery dealerships in the late 1980s when eight models were still available and pre-delivery inspections were carried out at the dealer's premises in Essex, Shropshire and Suffolk.

The three-cylinder Ursus 2812 and four-cylinder 3512 and 4512 made at the new factory were among the 35–150 hp range of Polish tractors imported in 1991 by BSG Supplies Ltd, originally at Ipswich but later at Colchester in Essex.

315. The Ursus Lightforce de luxe range introduced in 1996 included the 62 hp 4512.

Seven different Ursus tractors from the 35 hp 2812 to the 150 hp 1224T were new models with improved styling and most had hydrostatic steering. The front axle diff-lock was hydraulically engaged via the rear diff-lock but the hydraulically engaged four-wheel drive was an optional extra.

Production difficulties arising from the political situation in Eastern Europe caused the Ursus factory to close for a while in 1991. However the Polish tractors soon reappeared and eighteen two- and four-wheel drive tractors, including the first of the new Lightforce range and restyled heavy tractors with a Zetor connection, were introduced in time for the 1992 Royal Smithfield Show. The two-wheel drive tractors ranged from the 35 hp Ursus 335 to the 89 hp 1222 and the four-wheel drive models were in the 70–119 hp bracket.

The twin-cylinder 335 was mechanically much the same as it was in the mid-1960s and like the three-cylinder 37 hp 2812 and 47 hp 3512 the Ursus 335 had an eight forward and two reverse gearbox. The 912, 1012 and 1222 rated at 78, 100 and 119 hp had a sixteen forward and eight reverse gearbox but the four-wheel drive models only had four reverse gears. There were three versions of the Ursus 4512 with an eight forward

and two reverse gearbox. The standard 4512 and four-wheel drive 4514 had a 70 hp engine and the turbocharged 4514T was rated at 80 hp.

The Ursus Mechanical Works celebrated its centenary in 1993 with the launch of a heavyweight range of tractors with a synchromesh gearbox and an adjustable steering column in improved cabs.

An independent power take-off and live hydraulics were features of the 37–90 hp Lightforce HD models introduced in 1996 and the turbocharged two- and four-wheel drive Ursus 3512/3514 rated at 62 hp were added to the Lightforce range in the same year. BSG Supplies were still importing five Ursus models in the 38–86 hp bracket in 1997, when standard equipment of these two- and four-wheel drive tractors included three- or four-cylinder engines, an eight forward and two reverse gearbox, diff-lock and hydraulic linkage.

Valmet

Valmet tractors took their name from the Valtion Metallitehtaat (State Metal Works) established in Finland in the late 1940s. A prototype single-cylinder 12 hp Valmet tractor was made in 1949 but it failed to come

316. Launched in 1964 the 52 hp Valmet 565 had a synchromesh transmission. (Stuart Gibbard)

up to expectations so the Valtion Metallitehtaat designers went back to the drawing board. The company became the Valmet Corporation in 1951 and a limited number of 15 hp Valmet 15, or Model A tractors, with a four-cylinder side-valve vaporising oil engine and magneto ignition were made in 1952. Finland's farmers wanted more powerful tractors and this demand was met with the launch of the more popular Valmet 20 with a 22 hp petrol or 19½ hp vaporising oil engine in 1955. Hydraulic linkage was optional for a while but soon became a standard feature.

Valmet tractors were popular in Finland by the mid-1950s when the Finnish company was building up export markets in Brazil, China, Spain and Turkey. Valmet were also developing their own diesel engine and the 37 hp 33D tractor with a three-cylinder direct injection diesel engine, a six forward and two reverse gearbox and hydraulic linkage was launched in 1956. The Valmet 359D with the same 37 hp engine replaced the 33D in 1959 when improvements included an engine-driven hydraulic pump with a separate oil circuit and additional dials and gauges on the instrument panel.

Valmet formed a marketing organisation with Finnish farm machinery manufacturers Fiskars, Rosenlew and Wartsilla in 1960. As Brazil was an important market

Valmet opened a factory there in the same year in order to build the Valmet 360D tractor. With the exception of its German MWM (Motoren Werke Mannheim) diesel engine, the 360D tractor was identical to the Finnish-built 359D. The more stylish-looking 42 hp Valmet 361D, which replaced the 360D in 1960, was updated in 1962 with the addition of a diff-lock, hour meter and cushioned seat. The Valmet 363D articulated tractor was added to the range in 1963. The Valmet 361D which remained in production until 1964 was also built under licence in Portugal during the mid- and late 1960s and sold in that country as the FAP 361D (Fábrica de Automóveis Portugueses) 361D. About 700 FAP 361D tractors were sold in Portugal, Spain and Portuguese Africa.

The 52 hp Valmet 565 with a six forward and two reverse semi-synchromesh gearbox, foot throttle and a dual clutch with two equal-sized clutch discs and separate pedals replaced the 361D in 1964 and the 80 hp 864D was launched the same year. The engine-mounted hydraulic pump on the improved Valmet 565-II introduced in 1965 was modified so that it could be disengaged to reduce the load on the engine when starting from cold. The four-cylinder 90 hp Valmet 900 with an integral safety cab introduced in 1967 had an eight forward and two reverse synchromesh gearbox, a

540/1,000 ground speed power take-off, hydrostatic steering and hydraulic dry disc brakes. Yellow and brown paintwork replaced the previous red livery and the instrument panel and gear levers on the right-hand side of the seat in the safety cab would have been equally at home in a motor car. The new colour scheme was also used for the Valmet 700 and for the Valmet 500, which replaced the 565 II in 1968. The first two figures of the model number indicated engine horsepower rounded to the nearest ten. The 75 SAE hp Valmet 700 was mechanically similar to the Valmet 500 and both tractors had disc brakes, an improved eight forward and two reverse synchromesh gearbox and wider mudguards.

317. The 75 hp Valmet 702, which superseded the 700 in 1972, had hydraulically operated clutch and brakes with suspended pedals in the 'quietest cab yet made'.

The 102 hp four-cylinder Valmet 1100 announced in 1969 with a similar specification to the 900 was claimed to be the world's first turbocharged tractor. The four-wheel drive 900 and 1100 tractors were popular with agricultural contractors in Scandinavia. The Valmet 1100 was also advertised as the first European tractor to have in-cab entertainment (radio) included in the basic price.

Volvo BM also introduced a tractor with a six-cylinder turbocharged engine at roughly the same time and the argument concerning who had made the first turbocharged tractor was no longer an issue when Valmet and Volvo BM pooled their resources in 1979.

The Valtra name, derived from Valmet Tractors, came into in use in the early 1970s for a range of Valmet industrial tractors, fork lifts, cranes, loaders and other equipment but the Valtra name did not appear on farm tractors until 1982.

The Valmet 500 with a new flat floor quiet cab became the 502 in 1972. Similar flat floor cabs were used on other Valmet tractors over the next two years when the model numbers were changed to the 702, 902 and 1102. The turbocharged 102 SAE hp Valmet 702S appeared in

1975 and four-wheel drive versions of the 702/702S and a new 58 hp Valmet 602 were launched in 1978.

The 150 hp Valmet 1502 bogie tractor introduced in 1975 had six equal-sized wheels that were smaller than the usual rear wheels. It was conventionally steered by the front wheels and rubber tracks could be fitted over each pair of rear driving wheels to reduce ground pressure. The 1502 was popular in forestry and other industrial applications but it was not successful as a farm tractor.

A new brighter yellow and chocolate brown colour scheme was used for the 02 series from 1978 and for the new Valmet 03 series. Quick-attach couplers on the hydraulic lift arms and top link were introduced in the same year. The Valmet 703 had a synchronised speed splitter and a new 107 hp engine was used for the Valmet 1103.

Valmet and Volvo signed an agreement of co-operation and formed a joint company called Scantrac AB in 1979 and thirteen new Valmet tractors in the 58–136 hp bracket, six with four-wheel drive and seven with turbocharged engines, were launched in the same year.

The Valmet 04 series and the Volvo BM Valmet 05 series Nordic tractors with red and black paintwork and white wheels were launched in 1982. Although the 04 series were badged as Valmet tractors they were part of the new Volvo BM Valmet Nordic tractor range. The

Valmet 49 hp 504 and 61 hp 604 with a restyled wedge-shaped bonnet replaced the 02 series and the two-wheel drive 604T and four-wheel drive 604 T-4 had turbocharged engines.

The Volvo BM Valmet 05 series were the most powerful tractors in the Nordic range. The 505, 605, 705 and 805 with 65, 72, 83 and 94 hp engines respectively had a synchromesh gearbox and the planetary change-on-the-go Trac-Trol system with sixteen forward and eight reverse gears. Four-wheel drive versions of the 605, 705 and 805 were also made.

Bamfords of Uttoxeter sold Volvo BM tractors in Britain for a short period in the late 1960s and the Scandinavian tractors reappeared again in 1984 when Nordic Tractors at Thetford in Norfolk introduced the 05 series to British farmers. Prices ranged from £12,500 for the two-wheel drive 505 rising to £25,000 for the four-wheel drive 805. An advertisement, which suggested farmers should try these Volvo BM Valmet tractors for themselves, carried the slogan, 'When you've driven one you'll want to own one – and when you own one you'll want to drive one.' Improved cabs with optional air conditioning were used on the 05 series when the two- and four-wheel drive 53 hp 305 and 61 hp 405 along with the four-wheel drive 905, 2005-4 and the 2105-4

were launched in 1985. A 105 hp naturally aspirated six-cylinder engine was used on the 905, the 2005-4 had a 140 hp turbocharged power unit and the 2105-4 had a 163 hp turbocharged and intercooled engine.

When Valmet acquired Volvo BM's share of Scantrac AB in 1985 they were no longer able to use the prestigious Volvo BM name. The two- and four-wheel drive 53 hp Valmet 305 and 61 hp 405 launched in 1985 were exhibited on the Nordic Tractors stand at that year's Royal Smithfield Show. In common with the other 05 series tractors the engine and gearbox were mounted well forward to leave space for the fuel tank and reduce the need for front weights. The design also made it possible for an engineer to split the tractor for repair purposes without removing the cab.

Scantrac UK at Bury St Edmunds, a subsidiary of Scantrac AB Sweden, replaced Nordic Tractors as the Valmet distributor in 1986. They showed great faith in Valmet tractors by using the slogan 'Valmet – a legend in its own lifetime from the land of the Gods and Vikings' on their advertisements for the 53–163 hp Valmet 04 and 05 series tractors. Valmet Tractors AB were selling the 05 series tractors from their depot in North Yorkshire in 1987 and the Benson Group marketed the Finnish tractors from 1989 until 1991 when the Finnish

319. Farmers had the choice of five colours for the bonnet and cab on the Valmet 05 series tractors.

tractor manufacturer established Valmet Tractors UK at Runcorn in Cheshire.

Improved models of the Valmet 505 to 905 Power Plus tractors introduced in 1988 had 72–110 hp engines, a new front-wheel drive system, more efficient steering and the optional Autocontrol electronic hydraulic linkage control system used to pre-programme draft control settings. Farmers were also given the choice of red, yellow, white, green or blue paintwork for the bonnet and cab.

The 140 hp Valmet Mega 8300 and 170 hp Mega 8600, launched in 1989, were the first of the new high horsepower four-wheel drive Mega 8000 series. The 120 hp Mega 8100 was added in 1990. The specification of the new tractors included 40 kph transmissions with a twin-disc clutch, a sixteen-speed synchro Trac-Trol quick-shift system and the Valmet Autocontrol II electronic management system.

The 1991 Valmet range of 61–170 hp tractors included the new two- and four-wheel drive 79–120 hp Mezzo models launched at that year's Royal Smithfield Show. The new models had a twelve or twenty-four speed forward and reverse shift-on-the-move gearbox with an optional Hi-Trol fluid drive coupling installed in front of the main dry plate clutch to protect it from excessive wear. The Mezzo E versions had the three-way Autocontrol system for the diff-lock, the change-on-the-

move transmission and power take-off overload mechanism and the computerised Agridata system which provided full operational information. An advertisement explained that Autocontrol and Agridata would allow the driver 'to concentrate on his task with the minimum of stress'. The 140 hp Valmet 8400 with Delta Powershift transmission and the TwinTrac reverse drive was added to the Mezzo range in 1993.

Four new Valmet 55 series tractors also with a choice of five colour schemes appeared in 1992. The 61, 67, 72 and 80 hp Valmet 355, 455, 555-4 and 665-4 had an eight forward and four reverse synchromesh gearbox or sixteen forward and eight reverse gears with a shuttle reverser for the lower gears and an optional AutoControl computerised management system. A 75 hp Valmet 565 was also made but very few were sold.

The 61–75 hp Valmet M5 tractors for stock farms and the 190 hp Mega 8800 were launched in 1994. The M5 range included the 365-2/365-4 and 465-2/465-4 with three-cylinder engines and there was a choice of a naturally aspirated or a turbocharged four-cylinder engine for the M565-4. An eight forward and four reverse gearbox was standard but an optional twelve forward and eight reverse gearbox with very low creep speeds was also available. AutoControl II was standard on the Mega 8800 and the improved AutoControl IV was available at extra cost. Sales leaflets explained that

320. The Valmet 6400 had a wastegate turbocharger, air-suspended seat and a wash wipe facility for the rear screen.

AutoControl IV 'changes gear, engages four-wheel drive, monitors wheelslip, engages the diff-lock, monitors engine temperature, adjusts fuelling and does everything else the driver needs for day-long efficient operation'.

Valmet was still a state-owned company in 1994 when it merged with the similarly state-owned Sisu Machinery, which at the time was making trucks, military vehicles and diesel engines. Sisu also distributed Valmet M5, Mezzo and Mega tractors in the UK from their depot at Runcorn in Cheshire. The Valmet Sigma Power Mega 8750 introduced in 1995, advertised as a tractor with a split personality was described as 'a compact and lightweight 160/190 hp model that becomes a real heavyweight when the driver turns up the power'. The Sigma Power computer monitored the load at the power take-off and automatically increased the fuel supply when the power take-off shaft absorbed more than 30 hp. This additional power was divided between the transmission and power take-off.

Four new Valmet Mega 50 tractors in the 110–160 hp bracket replaced the Mega 70 series in 1996 and the 160/190 hp Sigma Power system was optional for the 160 hp Mega 8750. Sisu Machinery Ltd sold the Valmet

tractor business to Partek in Finland in 1997 and the two- and four-wheel drive Valmet Mezzo 6200 with a four-cylinder 80 hp engine was introduced in the same year. The new owners formed Valtra Tractors (UK) Ltd and sixteen Valtra-Valmet models in the 60–190 hp bracket with a red, black and white colour scheme were on the UK market in 1998. They included four 60–90 hp M100 tractors, replacements for the M5 series and the Mezzo and Mega ranges from the 80 hp Mezzo 6200 to the 160/190 hp Mega 8750 Sigma Power. The new 80–200 hp Valtra Valmet HiTech range launched in 1998 had four- and six-cylinder engines, automatic four-wheel drive, programmable powershift transmissions, air-suspended front axles and the flagship 160/200 hp SigmaPower 8950 developed its maximum power when doing heavy power take-off work. The Valmet 8350 EcoPower, which developed its full 135 hp at 1,800 engine rpm and the improved HiTech 2.0 tractors with a programmable powershift transmission appeared in 2000.

Valmet celebrated fifty years of tractor production in 2001 when they introduced the new Valtra name and logo and launched the 100–120 hp X Series articulated tractors for farm and forestry work in the same year. The

321. The first Valtra S series tractors had a six-cylinder engine and a forty forward and reverse speed transmission.

first Valtra S series tractor and the 120–190 hp Valtra T series that replaced the 8000 HiTech range were added in 2002. The six-cylinder 260 hp S series model had a forty forward and reverse speed power shuttle transmission and the 120–209 hp Valtra T series tractors had six-cylinder Sisu engines. Features of the T series included a thirty-six speed forward and reverse power shuttle transmission, adjustable load-sensing hydraulic control valves in the cab and new styling included a one-piece bonnet. The flagship T190 Sigma Power model had an electronically controlled engine which, depending on the work being done, developed 174, 189 or 209 hp. The nominally rated 174 hp increased to 189 hp in second and third powershift gears and the engine provided 209 hp when the power take-off was doing heavy work.

The 2003 Valtra tractor range included the new C, M, XM, S and T series tractors, the 63–98 hp A series and the 80, 90 and 100 hp Mezzo range with Sisu turbocharged four-cylinder engines and synchroshuttle transmission. The two most powerful tractors in the new seven-model 100–147 hp Valtra C series range had EcoPower engines which reached maximum horsepower at 1,800 rpm instead of the more usual 2,200 rpm and there was a choice of manual or automatic selection for four-wheel drive and the diff-lock. The four-cylinder 115–147 hp M series with electronic engine management had a thirty-six forward and reverse powershift transmission. The 130 and 147 hp Valtra XM with articulated chassis steering had the

same M series powershift transmission. Valtra sales literature claimed that the 147 hp Valtra M150 was the biggest four-cylinder tractor in the world. The TwinTrac reverse drive facility with a seat and driving controls which could be rotated through 180 degrees was an optional extra for the Mezzo, S series and T series tractors. Valtra became part of the AGCO Corporation in 2004.

Versatile

The Hydraulic Engineering Co founded in Toronto in 1947 used the Versatile name for their crop spraying machinery. The Toronto factory was too far away from their customer base so the company moved to a larger factory at Winnipeg in 1950 and made their first self-propelled grain swathers there in 1954. The Hydraulic Engineering Co changed their name to the Versatile Manufacturing Co in 1963 and the first Versatile four-wheel drive articulated tractor made its debut in 1966. Versatile was a leading manufacturer of high horsepower tractors in the mid-1970s when four models in the 230–350 hp bracket were on sale in North America. The 300 hp Versatile 900 had a Cummins V-8 diesel engine, a twelve-speed gearbox and articulated steering with two hydraulic rams supplied with oil by a 24 gallons per minute gear pump.

The company changed its name to the Versatile Farm Equipment Co in 1977 when they introduced the revolutionary four-wheel drive Versatile 150 Bi-directional tractor with a 71 hp Perkins engine, which worked equally as well going forwards or backwards and the operator always faced the direction of travel. It was not a new concept as Deutz had made a bi-directional ploughing tractor way back in 1907 but the Versatile 150 driver had the added benefit of four-wheel drive, hydraulic linkage and a cab. Later Bi-directional drive models included the 86 hp 160 launched in 1981 followed in 1986 by the 256 with a 100 hp Cummins engine. The even more powerful 276 with a 116 hp Cummins engine was introduced in 1985. The Bi-directional Versatile 276 had a front and rear power take-off and hydraulic linkage and when it was used with a front-mounted cutter bar it became a high-output self-propelled swather.

An agreement with Fiat Trattori of Turin gave Versatile an entry into Europe when the Fiat 44-28 with a 'Constant-Power' six-cylinder 280 hp Cummins engine was introduced to UK farmers at the 1979 Royal Smithfield Show. It was one of four 230–350 hp Versatile models in production at the time and the 'Constant-Power' feature gave the Fiat 44-28 a constant 280 hp from the Cummins engine through a speed range of 1600–2100 rpm.

The Versatile Farm Equipment Co met the North American prairie farmers' demands in the early 1980s for even more horsepower. The 1976 600 hp Versatile Model 1080, known as Big Roy, was the flagship model in a range of 230 hp-plus giant tractors. The eight-wheel drive articulated Big Roy had four axles and the Cummins engine mounted high up behind the cab virtually covered the rear window. A closed-circuit television system in the cab enabled the driver to see what was happening behind Big Roy's back.

Described in the farming press in 1982 as the 'biggest tractor in the UK' the 20 ton 470 hp Versatile 1150 with an £83,000 price tag was shown in the UK by Fiat Agriculture. Badged as the Fiat 44-47 it had a six-cylinder turbocharged and after-cooled engine, an eight forward and four reverse constant mesh gearbox and a hydraulic system with five spool valves. The 44-47's twin fuel tanks held enough diesel to keep the engine running for sixteen hours without refuelling, a ladder at each side of the engine provided access for servicing and the air-conditioned cab with tinted glass had a tilting and telescopic steering column.

Ford New Holland bought the Versatile Farm Equipment Co in 1987 after Steiger, which made the Ford FW series tractors, had been acquired by Case IH in 1986. Ford New Holland's acquisition of Versatile provided them with a new source of high

322. The Versatile 876 was one of the first four Canadian-built high horsepower tractors with a New Holland badge.

323. The articulated 325 hp Ford 946 was made at the Versatile factory.

horsepower tractors and the Versatile 276 in Ford blue livery was introduced to British farmers in 1988 to test their reaction to the big centre-pivot steered tractor with a three-range hydrostatic transmission.

The 325 hp turbocharged Ford Versatile 946 articulated tractor with twelve forward and two reverse gears, added to the Ford New Holland range in the late 1980s, was still being made in 1993. Although very high horsepower tractors were less popular in the late 1990s the 1997 New Holland price list still included the 360 hp Versatile 9682. The Bi-directional New Holland TV140 with a six-cylinder 115 hp engine and a three-speed hydrostatic drive was launched in 1998. However, following the Case New Holland merger in 1999 the Versatile name, production rights and factory were divested to Buhler Industries at Winnipeg in 2000 and the new owners introduced a range of 290–535 hp Versatile tractors.

324. The supercharged Rolls Royce engine on the Vickers VR180 Vigor developed 180 hp. (Stuart Gibbard)

Vickers

Vickers-Armstrong (Tractors) Ltd at London and Newcastle-upon-Tyne made the largest British-built crawler tractors in the 1950s. Jack Olding & Co at Hatfield in Hertfordshire, who had previously marketed Caterpillar tractors in the UK, distributed two models of Vickers crawler with Rolls Royce engines. The big crawler tractors were designed for the construction industry but in an attempt to sell them to farmers the tractors were also advertised in a number of farming magazines.

The 1952 Vickers VR180 Vigor crawler with a Rolls Royce six-cylinder supercharged direct injection diesel engine and a six forward and three reverse gearbox had a torque converter which allowed the driver to change gear on the move. It was steered by a system of servo-actuated multi-disc clutches and manually operated band brakes. The Vigor crawler, which weighed almost 15 tons, developed 181 hp at the power take-off when it was tested by the National Institute of Agricultural Engineering (NIAE) in 1953.

A pre-production model of the smaller 140 hp Vickers Vikon crawler launched in 1956 was submitted to the NIAE for testing. The four-cylinder four-stroke diesel crawler with clutch and brake steering had a single-plate clutch and a five forward and four reverse gearbox with a top speed of 5½ mph. The Vikon had front and rear power take-off shafts and the tractor, including the driver and a full tank of fuel, weighed 11½ tons. The last Vickers tractors were made in the early 1960s.

Volvo BM

The Swedish vehicle manufacturer AB Volvo was established in 1927 and they made the first Volvo tractors in 1943. The BM side of the company can be traced back to the Swedish Bolinder and Munktell companies of the mid-1800s. The Bolinder brothers, who started an engineering and foundry business in Stockholm in 1844, were making internal combustion engines by the mid-1890s. Johan Theofron Munktell made a printing machine in 1830 and built the first Swedish steam railway engine in 1853. He was making stationary steam engines in the mid-1850s and Munktell threshing machines in 1859.

The first Munktell tractor with 6 ft diameter rear wheels and weighing over 8 tons appeared in 1913. It had two 15 hp single-cylinder two-stroke hot bulb engines, which coupled together developed about 30 hp. The cooling system held about 90 gallons of water. Sweden had to import all of its oil products so Munktell experimented with wood-burning engines for his tractors. A smaller single-cylinder Munktell 22 made between 1921 and 1934 weighed just over 2 tons and Munktell tractors took part in the 1930 World Tractor Trials in Oxfordshire. Bolinder and Munktell joined forces in 1932 as BM Bolinder-Munktell and made their first hot-bulb BM20 tractors with electric starting in 1944. The 20 hp BM10 with the radiator mounted sideways behind the engine and the fuel tank and between the engine and gearbox was added in 1947. AB Volvo acquired Bolinder-Munktell in 1950 to form Volvo BM.

The BM10 and BM20 were discontinued in 1952. The new 43 hp BM21, which was the last BM tractor with a hot bulb engine, appeared in the same year. The 40 hp BM35 and BM36, also introduced in 1952, were the first BM tractors with direct injection diesel engines.

The BM55 and Volvo T55 tractors with 50 hp four-cylinder diesel engines launched in 1954 were identical apart from their paintwork. The only difference between the BM230 Victor and the Volvo T230 introduced in 1955 with a 29 hp twin-cylinder diesel engine and five forward and one reverse gearbox was the colour scheme, the BM model was green with an orange engine and the T230 was red. Improvements were made to the gearbox and hydraulic system in 1958 and from 1960 both versions had red paintwork. About 16,000 Victors 230 and T230 tractors were in use in many parts of Europe when it was discontinued in 1961 but none were sold in the UK.

Volvo BM introduced the 425 Terrier with a 32 hp petrol engine in 1956 and when a 40 hp three-cylinder Perkins diesel replaced the petrol engine in 1961 the Terrier became the BM Buster 320A. The three-cylinder 56 hp BM 350 Boxer with ten forward and two reverse gears appeared in 1959 and the 79 hp four-cylinder BM Bison was added in the early

1960s. With a more powerful 47 hp engine the Buster became the Buster T400 in the mid-1960s. The 106 hp six-cylinder T800 superseded the Bison in 1966, the Boxer with a 66 SAE hp engine became the T600 in 1968. The T810 and the four-wheel drive T814 with 130 hp turbocharged engines, which were improved versions of the earlier T800, appeared in 1969.

More new models including the very successful Volvo BM T650 were launched in the early 1970s. The BM T650 was a completely new tractor with an 80 SAE hp four-cylinder engine, eight forward and two reverse gears and a cab of saloon car quality. Although the standard gearbox did not have synchromesh Volvo's change-on-

325. Introduced in 1956 the BM 425 Terrier had a 32 hp petrol engine.

326. The diesel-engined BM Volvo Buster 320 superseded the 425 Terrier in 1961.

327. Bamfords of Uttoxeter distributed Volvo BM tractors including this T800 with a Bamford plough in the UK in the late 1960s.

the-move Tract-Tol gearbox was later added as an optional extra. The T500 introduced in 1974 had a Perkins engine and an International Harvester transmission made at Doncaster. More models were added and by 1978 Volvo BM were making a comprehensive range of two- and four-wheel drive tractors.

Plans were well advanced for the new Nordic 04 and 05 series in 1979 when an agreement of co-operation between Volvo BM and Valmet resulted in the formation of Scantrac AB. Launched in 1982 the new 05 series included the 605-4, 705-4 and 805-4 with 72 hp, 83 and 95 hp engines and a sixteen forward and eight reverse transmission. The Volvo BM Valmet four-wheel drive 905-4 was added later in the year. Its 150 hp six-cylinder engine installed well forward to place almost half of the engine weight over the front axle had a sixteen forward and eight reverse gearbox. The two- and four-wheel drive Volvo BM Valmet 53 hp 305 and 61 hp 405 were added in 1985 when Valmet bought the AB Volvo share of Scantrac AB. From that date Valmet were no longer able to use the Volvo name on their farm tractors.

Wheels and Tracks

The change from steel wheels to self-cleaning open-centre tread pneumatic tyres took the best part of twenty years. Smooth pneumatic tyres without any tread were used on farm tractors for the first time in 1931 and tyres with specialist agricultural tread patterns came into use in the late 1930s. A shortage of rubber during the war years slowed the change to pneumatic tyres and they were not standard on farm tractors until the mid-1940s. Some farmers still used spade lug steel wheels for

328. This Fordson Model N with a mechanical implement lift system has closed centre tyre treads.

ploughing and other heavy work in the late 1940s when it was still possible to buy a new E27N Fordson Major and other tractors with steel wheels.

Various joined-up tread patterns were used before the self-cleaning open-centre tread came into universal use. A number of companies including Blue Peter and Vacu-Lug, established in the 1950s, specialised in re-treading rear tractor tyres. Farmers using the service were told that their own tyres would be returned to them as good as new.

EA Allman & Co at Chichester was one of a number of companies making various types of tractor wheel in

329. Most tractors had self-cleaning open centre tread tyres by late 1940s.

the late 1940s. The Allman range included complete sets of wheels and tyres to convert a tractor from spade lugs to pneumatic tyres, skeleton steel wheels and narrow wheels with open-centre tread tyres for rowcrop work.

Spade lug and skeleton wheel rims, which were bolted to the standard dished wheel centres, were still included in the 1950 list of accessories for the Ferguson TE20 tractor. Spade lug wheels had a number of steel lugs bolted on a wide rim and skeleton wheels for rowcrop work had detachable lugs bolted alternately to the inside and outside of the rim or on only one side when working in closely spaced rows. Some farmers bought complete wheels but others saved money by using rowcrop or skeleton rims bolted on to the rear wheel discs after the rims and tyres were removed. Wheel bands had to be fitted over the spade lugs before taking the tractor on the public highway.

330. With road bands over the spade lug wheels this Fordson tractor was ready for the road.

Strakes, girdles and tyre chains used to improve wheel grip in greasy conditions were popular in

331. Some farmers used Miller skeleton steel wheels for rowcrop work.

332. A Stanhay Darvill strake on the rear wheel which ran on the unbroken land improved grip when ploughing.

333. Tractors with steel tracks including the 1940 Caterpillar D2, 1951 New Fordson Major on County full tracks and 1948 County E27N working at the Tracks Across the Field event had more pulling power than the equivalent wheeled tractors.

the late 1940s and 1950s. The heavy Opperman 'Quickgrip' strakes bolted to the tractor rear wheel discs had hinged lugs which were folded over the tyres to improve wheel grip. Darvill retractable strakes made by Stanhay at Ashford in Kent, which were bolted to the rear wheel discs, were easier to use. The spikes were extended beyond the tyres by loosening two lock nuts and driving the tractor forward. The procedure was reversed to close the strakes before taking the tractor on the road. A strake on the rear wheel which ran on the unbroken ground was an advantage when ploughing but the combined width of the strake and the tyre was too big to run in the furrow when reversible ploughing. With the arrival of improved hydraulic weight transfer systems wheel strakes were left in the tractor shed.

Some farmers used chains or tyre girdles over the rear wheels to improve traction in greasy soil conditions. Griff non-skid tyre chains wrapped around the tyre were claimed to give 'a sure, safe grip to all tyres however old and smooth in the field and on the road without harming them'. Ferguson and other makes of tyre girdles, usually in five hinged sections, were wrapped around the tyres and could be used in the field or on the road without adjustment. They were designed to creep slightly round the tyre when under load to prevent damage and assist self-cleaning of the tyre tread. Spade

lugs could be bolted to the cross bars of the girdles for extra grip in the field.

Increased draw pull was achieved with steel half-tracks made by Roadless Traction. The tracks, carried on a drive sprocket attached to the rear axle shafts and a rear idler wheel, had a much larger footprint area than pneumatic tyres and consequently reduced soil compaction. Roadless sales literature suggested that a set of half-tracks could be fitted or removed in a few hours but it usually took much longer and few farmers ever bothered to try.

Rotaped tracks, introduced by Geo Monro Ltd at Waltham Cross in Hertfordshire in 1946, were an alternative method of converting a wheeled tractor to a half-track. Mainly used for drainage work Rotaped tracks were attached to the rear axles and the driving sprockets were bolted to the rear wheel hubs. The back of the tractor was carried on six track rails which were moved around the central driving sprocket by a system of diagonal chains. Advertisements for Rotaped tracks explained that they increased pulling power by up to 70 per cent, did not clog or slip and when required it was not difficult to put the tractor back on to its rear wheels. Leeford (London) Ltd announced a lighter type of Rotaped track suitable for some of the more powerful tractors on the market in the late 1950s. The manufacturers claimed that the new tracks were more versatile than the earlier design and could be used at speeds of up to 4½ mph for ploughing and other heavy fieldwork.

Soil compaction was another cause for concern in the early 1950s and various designs of cage wheel, which gave a larger footprint area, were made at the time. Some farmers attached cage wheels to the tractor wheels to help reduce soil damage when preparing seedbeds, others removed the rear wheels and ran the tractor on cage wheels. The steel tread bars on cage wheels were made with angle iron or round bar but most tractor drivers preferred the round bar design as angled iron bars threw up clouds of dust when the land was dry.

334. Roadless made DG half-track conversions for most popular tractors.

Water ballasting of tyres to improve traction by adding weight was developed in the early 1950s. Goodyear introduced their Solution 100 system which, with the aid of a special pump, made it possible to fill the inner tube

335. Rotaped tracks were mainly used for field drainage work.

336. Cage wheels reduced soil compaction when preparing seedbeds.

100 per cent full with water. Other companies suggested that a smaller amount of water, which could be added without special equipment, would be equally effective in improving traction. Calcium chloride was added to the water to protect it from frost. Tyre companies had problems convincing some of their customers that ethylene glycol anti-freeze solution for tractor radiators was not suitable for use in inner tubes.

Hedge-cutting machinery manufacturer J Foster & Co at North Cave in Yorkshire found that tractor tyres soon became damaged or worn when hedge cutting or clearing scrub and to overcome this problem introduced the Sabard sectional steel tyre cover in the 1950s. The tyre covers were made in six hinged sections with ⅛ in steel plate and sales literature advised that the cover could 'be fixed around the tyre in a matter of minutes'.

A fortunate few had an air compressor in the farm workshop or a tractor-powered pump in the 1940s and 1950s when most tractor drivers had to pump up their tractor tyres with a foot pump. The Schrader tyre pump with a length of hose and pressure gauge provided an easier way to inflate tractor tyres. A threaded connector was screwed into a sparking plug hole after removing one of the plugs and with the engine ticking over on the remaining cylinders the selected piston acted as a pump to inflate the tyres. Small pto-driven air compressors for pumping up tyres or paint spraying were also used in the early 1950s. Lawrence Edwards & Co made a 100 psi capacity pto-driven pump with 18 ft of hose and tyre pressure gauge which could be attached to the pto shaft in a matter of seconds. The Edwards pump, advertised at £12.10s, was claimed to 'put an end to back-breaking tyre pumping and can inflate a full set of four Fordson Dexta or Ferguson tyres in under six minutes'. Farmfitters Ltd of Gerrards Cross in Buckinghamshire made the Jumbo pto-driven pump for inflating tyres and pumping liquids. The makers suggested that with suitable accessories the Jumbo pump could be used not only to inflate tyres but also to spray fruit trees, spread liquid manure, transfer fuel from a barrel into the tractor tank and blow dust from the engine radiator.

Soil compaction caused by the increasing weight of farm tractors was a major problem in the late 1970s. Oversize tyres were of some help but much wider terra tyres with a large footprint area were even better. The large ground contact area provided by crawlers and half-tracks was the best way to reduce soil damage but steel-tracked tractors were slow. Rubber half-tracks wrapped round the rear wheel and a smaller wheel in front of the tyre were used in the days of the Ferguson 20. This idea reappeared in the in the early 1990s when Kitco rubber half-tracks arrived on the market. Suitable for most makes of tractor, they could be fitted by two people in under an hour. The tracks worked well in most conditions but there was a risk of them coming off when reversing the tractor or making sharp turns on very dry land.

The Silsoe Research Institute demonstrated their rubber half-track conversion at various field events in 1990. Test results found that there was a 25 per cent increase in performance when compared with a standard two-wheel drive tractor. The Silsoe suspended rubber track (SRT), suitable for two- and four-wheel drive tractors, was positively driven by a sprocket bolted to the rear wheel hub and a compensating system maintained constant track tension. A parallel link air-spring suspension system was claimed to give a comfortable ride. Richard Stocks Ltd of Wisbech in

337. Very wide low ground pressure tyres with a large footprint area limit the damage to soil structure.

338. These early 1990s rubber half-tracks could be used with worn tyres; they cost about £900 and were suitable for most popular tractors.

339. SRI half-tracks, designed by the Silsoe Research Institute, could be used at speeds of up to 15 mph.

340. The first Caterpillar with Mobil-trac rubber tracks appeared in the UK in 1989.

Cambridgeshire exhibited the SRT tracks at the 1991 Royal Smithfield Show but the design was not a commercial success.

Rubber-tracked crawlers were developed by Caterpillar in America and the first Caterpillar Challenger 65 tractor with high-speed Mobil-trac rubber tracks and a top speed of 18 mph was demonstrated to UK farmers in 1989. The Challenger 35 and 45 with triangular-shaped rubber tracks appeared in 1994. Claas marketed the rubber-tracked Challenger range with their green and white livery from 1997 until 2002 when Caterpillar withdrew from the agricultural market and production rights were sold to AGCO.

The rubber-tracked Track Marshall TM200 launched in 1990 with a 210 hp Cummins engine and a pneumatic suspension system enjoyed little success and was discontinued in 1994. John Deere introduced their rubber-tracked 8000T series in 1997 and the 360 hp Case IH Quadtrac with four independently suspended rubber track units built at the Steiger factory in America appeared the same year.

Winget

Slater & England Ltd at Gloucester, better known for Winget concrete mixers, made the Winget tractor with a 7¾ hp air-cooled Lister SR1 direct injection diesel engine in the early 1960s. The tractor, which used 1½ pints of fuel in an hour, had a three forward and one reverse speed Newage gearbox, diff-lock, power take-off, adjustable wheel track and live hydraulic linkage with a Dowty pump driven from the engine crankshaft.

Three versions of the Winget 42 with a top speed of 7 mph introduced in 1966 were advertised as 'easy to handle and just as easy to buy'. The Winget 42 had the same 7¾ hp Lister SR1 single-cylinder air-cooled diesel engine, a three forward and one reverse speed gearbox

combined with the rear axle and an adjustable track width. The Series I Winget 42 with a hand-lift toolbar cost £483.10s, the Series II with hydraulic linkage was £492 and the Series III complete with a front-end loader was £560. About 500 Winget tractors had been made by Slater & England when the production rights and remaining stock of parts were sold to MSW Machinery at Wood Vale in London.

The new owners used the existing stock of parts to build a similar MSW ride-on tractor until the early 1970s when they introduced the MSW 550 four-wheel drive tractor with a 9 hp Lister SR1 diesel engine and live hydraulics. There were four models of the MSW tractor in 1973. The MSW 550M had a manual gearbox and the MSW 550HC with a Petter diesel engine had a Marshallmatic fully variable forward and reverse transmission. The 9 hp MSW 663M with a Lister SR1 diesel engine also had a manual transmission and the MSW 633HC with a Lister ST1 diesel engine rated at 11 hp had a fully variable Sidestrand 18 hydrostatic transmission. Variants of the 633 had a heavy-duty rear axle and optional Kohler petrol or Hatz diesel engines

341. A 1967 advertisement for the Winget tractor explained that it was 'easy to handle and easy to buy'.

were also available. Added in 1975, the 19 hp MSW 1250M and 1250HC with Lister SR2 diesel engines had manual and Sidestrand hydrostatic transmissions respectively.

Zetor

Skoda, Praga and Wikov were all making tractors in Czechoslovakia in the years between the two world wars. Praga motor ploughs included a lightweight 10 hp model and a much larger five-furrow model with the same 40–50 hp petrol engine as the Praga Grand motor car. The Wikov tractor engine was unusual in that it had separate carburettors for petrol and vaporising oil and the rear wheels were belt driven.

A factory opened at Brno in 1942 to make aircraft parts and V2 rockets was bombed in 1944 and the state-owned Zbrojovka Works took possession of the partly destroyed factory. The first two-cylinder water-cooled 25 hp Zetor 25 tractors with six forward gears and one reverse were made there in November 1945 and quantity production began in March 1946. The Zetor 15 was added in 1947 but it was decided to develop the

Zetor 25 and the smaller tractor was only made for a year or so. About 8,500 Zetor 25 tractors had been made by 1950 when production was moved to the nearby Works of Precise Engineering. It became the Zetor Works in 1952 and Zetor was still the only tractor manufacturer in Czechoslovakia in 1960. About 160,000 model 25 tractors had been made at Brno when the new Unified range 1 Zetor 3011 and 4011 were launched in 1962 followed by the 2011 in 1963.

The first Zetor tractors sold in the UK, in 1965, were imported through Sheerness in Kent by the Pride & Clarke Motor Cycle Co and delivered with a full toolkit and supply of spare parts! Within a year the Unified Range I Zetor tractors, introduced in 1960, were marketed in the UK by the Motokov Foreign Trade Corporation which had links with Skoda (GB) Ltd. The London-based import company advertised the Zetor 2011, 3011 and 4011 in 1967 for £585, £650 and £795 respectively. The Zetor 3013, a narrow version of the 3011, cost £670 and the four-wheel drive 3045 was £779. The tractors had two-, three- and four-cylinder direct injection diesel engines rated at 23½, 32½ and 47 SAE

342. Optional equipment for the Zetor 3011 included an air compressor with a tyre inflator, a front power take-off and a seat for the co-driver.

343. The Zetor Crystal 8011 was introduced to British farmers at the 1971 Royal Smithfield Show.

hp, a dual clutch, ten forward and two reverse gears, diff-lock, expanding shoe brakes, live power take-off and category II Zetormatic hydraulics with draft control. Zetor 3011 sales literature explained that the design of the tractor incorporated everything possible at the time 'to afford the driver appropriate comfort, facilitate control and ensure the highest possible efficiency'. Instruments were 'neatly arranged in the driver's full visual field' and a spring suspension seat could be adjusted with three positions available 'for drivers of different stature'. Variants of the Unified Range I models included the 2023 crawler, the four-wheel drive 3045 and 4045 and the half-track 3016 and 4016 with slatted rubber belts running round the rear tyres and central rubber-tyred bogie wheel.

The 5511 with new styling was added to the Brno range in 1966. It was the first Zetor tractor to have suspension springs on the stub axles that could be locked when required. The earlier 2011, 3011 and 4011 models were improved and restyled as the 2511, 3511 and 4511 in 1967. The 80 hp Zetor Crystal 8011 and four-wheel drive 8045 introduced at the 1971 Royal Smithfield Show shared many parts with the Ursus C-385. Both tractors were the result of a joint Czechoslovakian and Polish design and production

programme with the engines made by Zetor and the transmissions by Ursus. The Crystal was the first of a group of more powerful Zetor tractors. The two- and four-wheel drive Crystal 12011/12045 rated at 120 hp appeared in 1974 and the Zetor high horsepower range was completed in 1983 with the launch of the Crystal 8111/8145, the 100 hp 10111/10145 and the 16045 with a six-cylinder 160hp engine.

The two-wheel drive Zetor 4712, 5711 and 6711 '7' series with 45, 56 and 65 hp engines appeared in 1972. Zetor were not able use the 4711 model number as it was the registered trademark of a perfume manufacturer so the 42 hp tractor was designated the Zetor 4712. Variants of the '7' series included the two-wheel drive 4718, 5718 and 6718 with a safety cab, the cab-less four-wheel drive 5745 and 6745 and the 5748 and 6748 with a safety cab.

The Zetor '9' series 4911, 5911/5945 and 6911/6945 with safety cabs replaced the '7' series in 1978 and when the improved 5011, 6011 and 7011 Zetor '10' series was launched in 1980 there were Czechoslovakian tractors in the 50–160 hp bracket on UK market. The more powerful four-wheel drive tractors from the 8045 upward had a torque multiplier and turbochargers were standard on the 12045 and 16045. Prices of the '10'

293

series ranged from £4,293 to £16,480 plus carriage from Kings Lynn. Power steering, trailer air brakes and a quiet cab with radio were included in the price. Press advertisements suggested that farmers buying a new Zetor '10' series tractor on easy terms would benefit from 'a painless low interest rate purchase, improved engines, reduced diesel costs and the bare minimum of maintenance costs'.

The new Zetor 5245 four-wheel drive version of the 50 hp three-cylinder water-cooled 5211 was launched at the 1984 Royal Smithfield Show. The existing Zetor Crystal and Unified I tractors were restyled and modified with an increased capacity hydraulic system and a better

344. The four-wheel drive Zetor 10145 had a de luxe cab with an adjustable suspension seat.

cab with a fully adjustable seat. The addition of the 5245 gave Zetor six two- and four-wheel drive UR I tractors from 50–120 hp on sale in the UK. The 5245 had a dual clutch, ten forward and four reverse gears with synchromesh on the fourth to fifth ratio and a two-speed power take-off. This improved specification brought the 5245 into line with its more expensive competitors but it still had a power-assisted steering system and drum brakes instead of the more up-to-date hydrostatic steering and disc brakes. The improved 60 hp 6211 and four-wheel drive 6245 had the same gearbox as the 5245, a self-adjusting clutch, an independent 540/1000 power take-off with an air-operated clutch and hydraulic drum brakes. The front wheels were driven by a central shaft from a transfer box to the front axle and drive could be engaged and disengaged on the move. The 60 hp 6211 and 6245 were given an optional power boost when a British Holset turbocharger was introduced in 1986 as a factory-fitted extra which increased the power output to 75 hp. Six two- and four-wheel drive Zetor tractors with 46–112 hp engines were still on the UK market in 1986. The Unified series had ten forward speeds and

two reverse and the Crystal models had a sixteen forward and eight reverse gearbox. The Zetor range had been extended to eighteen models in 1989 with tractors in approximate 10 hp steps from the 50 hp two-wheel drive 5211 to the four-wheel drive Zetor 14145 with a six-cylinder 140 hp engine.

The Zetor 9520/9540 introduced in 1990 was the first of a new range of turbocharged Czechoslovakian tractors. The 90 hp four-wheel drive tractors had an air compressor to operate the electro-pneumatically engaged diff-lock and trailer braking system, an airline was provided in the toolbox for inflating tyres and the compressed air supply could also be used for small power tools and paint spraying.

More new tractors were launched in 1992 and within a year there were ten models in the three Zetor Unified ranges with 50–140 hp power units. The three-cylinder 50 hp two- and four-wheel drive 3320/3340 were the smallest of the UR I series with 60, 70 and 84 hp four-cylinder models completing the group. Features of the UR I models included a ten forward and two reverse synchromesh gearbox, an engine-mounted hydraulic

345. A compressor and airline were standard on the Zetor 9540.

system pump, hydrostatic steering and multi-plate dry disc brakes. The UR II series consisted of two- and four-wheel drive 80 and 90 hp four-cylinder tractors and the four-wheel drive 11245, 12245 and 14245T with six-cylinder 105, 120 and 140 hp engines and a sixteen forward and eight reverse speed gearbox. The 90 hp turbocharged 9520/9540 was the only model in the UR III range in 1993 when prices ranged from £11,050 for the Zetor 3320 to £33,180 for the flagship 14245.

An agreement with John Deere in 1993 gave the American company distribution rights for the 40–85 hp Zetor tractors with green and yellow paint and a running deer badge. The deal also included the supply of diesel engines made at Brno for other John Deere equipment. The 'no-frills' Zetor 6340E economy version of the 6340 with a standard cab, hydrostatic steering and a compressor was added in 1994 and to avoid confusion the standard 6340 became the 7340.

The new UR I Super range with restyled cabs and bonnets was launched in 1998. The four-cylinder 60, 70 and 84 hp tractor model numbers were changed from 4320/4340 to 4321/4341 and so on and the 75 hp 6340E became the 6431ECON. Hydraulic lift capacity was increased and transmission options included a ten

forward and two reverse manual gearbox or a ten forward and reverse shuttle transmission.

The tractor and agricultural machinery division of Motokov UK at Kings Lynn still distributed Zetor tractors in 1998 including the UR III series with wet disc brakes, an air compressor and a choice of gearboxes. The two- and four-wheel drive 7520/40, 8520/40 and 9520/40 had four-cylinder engines rated at 70, 80 and 90 hp. The flagship four-wheel drive 1054 had a turbocharged and inter-cooled 105 hp engine and the 70 hp 7540E economy model had a no-frills specification and a price to match. In common with badge engineering elsewhere in the industry most of the Zetor models made at Brno were sold in certain parts of the world as John Deere tractors.

The four-cylinder 80–105 hp Zetor Forterra UR III series with a completely redesigned cab and interior layout was introduced to farmers at the 1998 Royal Smithfield Show. A new 110 hp six-cylinder UR III Forterra 11641 with a SAME Deutz-Fahr engine, a twenty-four forward and eighteen reverse speed powershift transmission and electronic linkage control hydraulics went into production in Czechoslovakia at the same time but it was not available in the UK.

346. The 75 hp four-wheel drive Zetor 7341 Super turbo was the largest model of the UR I range of tractors introduced in 1998.

Motokov UK imported a limited number of Zetor UR I Super and UR III Forterra models in 2000. There were four 50–80 hp UR1 Super two- and four-wheel drive Zetor tractors with a ten forward and two reverse gearbox and three 82–105 hp UR III models with a twenty-four forward and eighteen reverse synchroshuttle transmission. Motokov UK was also marketing EuroTrac, Landini and TYM tractors from Kings Lynn at the time.

Zetor added the UR I Major series to the existing UR I Super and introduced an improved range of UR 3 Forterra tractors in 2002. The UR I Major series included the Zetor 5340, 6340 and 7340 with four-cylinder 70, 75 and 84 hp engines and a ten forward and two reverse gearbox. The Forterra 9641, 10641 and 11441 models had four-cylinder 92, 103 and 110 hp engines and the flagship six-cylinder 11741

developed 120 hp under its bonnet. All four tractors had turbocharged engines and a twenty-four forward and eighteen reverse synchroshuttle transmission. Zetor launched the four-wheel drive Proxima range in 2004. Features included four-cylinder water-cooled engines, quieter cabs with the exhaust muffler under the bonnet, wet disc brakes and an air compressor for inflating tyres and the servo assisted the power take-off hand clutch for engine and ground speed power take-off.

The Proxima range with a Carraro front axle and four Forterra models were current when Zetor UK opened new premises at Downham Market in Norfolk in 2005. The 99–126 hp four-cylinder Forterra range had a synchro-shuttle twenty-four forward and eighteen reverse transmission with three powershifts and a full-size passenger seat in the cab.

Glossary

Badge engineering. Equipment made by one manufacturer and sold by another company with its own badge and colour scheme.

Belt horsepower. Power available at the belt pulley.

Brake horsepower. Power available at the engine flywheel.

Category I, II and III linkage. Standard dimension specifications used for tractor hydraulic three-point linkages. The Ferguson TE20 has category I linkage and very high horsepower tractors have category III linkage.

Centrifugal clutch. Two or three spring-loaded friction-lined shoes attached to the crankshaft rotate inside a hub on the gearbox input shaft. When the engine reaches a preset speed the shoes are thrown outwards centrifugally to grip the hub and transmit drive to the transmission.

Closed centre hydraulics. The pump only provides sufficient oil to carry out the required hydraulic functions; at other times the pump maintains the circuit pressure but reduces oil flow to save power.

Clutch and brake steering. Used for crawler tractors. The tracks are driven separately through a clutch unit with an integral brake for each track. Direction of travel is changed by using a lever to disengage partly or stop the drive to one track. Sharp turns are achieved with further movement of the lever that also applies the brake.

Cone clutch. Rotating male and female cones, usually with friction linings on the flywheel and gearbox input shaft, transmit power to the transmission. Separating the cones with a pedal or a lever disengages the drive. Field Marshall tractors have a cone clutch.

Controlled differential steering. An alternative steering mechanism for crawler tractors using an epicyclic gear system and a brake for each track. When the brake is applied one track slows down and the speed of the opposite track is increased.

DIN horsepower. The European standard used for nett engine power.

Donkey engine. A small petrol engine sometimes used to start diesel-engined crawler tractors.

Drawbar horsepower. Power available at the drawbar. It will always be lower than brake hp because energy is used for the transmission, hydraulics, etc.

Dual clutch. Two separate clutch discs are attached to the engine flywheel. The clutch pedal is pushed halfway down to disengage drive to the gearbox and fully down to stop the power take-off shaft.

Electronic Linkage Control (ELC). Electronic load sensing pins for the lower link arms measure the load on the tractor linkage and raise or lower the implement to maintain the pre-set level of draft.

Gasoline. The American term for petrol.

Ground speed power take-off. A power take-off shaft that turns at a speed directly related to the speed of the rear wheels irrespective of the selected gear ratio.

Horsepower. Horsepower is the measurement of the ability of a tractor to pull a load (lb) over a distance (feet) in one minute. Early calculations determined that one mechanical horsepower was to be 50 per cent more powerful than a carthorse. Various horsepower categories have been used for farm tractors. Belt pulley and drawbar horsepower were used in earlier times, for example the International 10–20 developed 10 hp at the drawbar and 20 hp at the belt pulley. Tractor power is now quoted in kilowatts and as 1 hp is equivalent to 746 watts, an engine developing 100 hp is also rated at 75 kW. The power of continental tractors is often given in cv, pf, pk, hk, etc which is 1 metric hp or 1.014 hp.

Hydrostatic steering. A hydraulically operated steering system with no mechanical linkage between the steering wheel and the front wheels.

Independent power take-off. Controlled by a separate clutch unit, usually operated hydraulically with a hand lever and not linked to the main transmission clutch.

Intercooler. This reduces the temperature of the air compressed by the turbocharger en route to the engine cylinders. As the air is cooled it becomes denser and increases the power output of the engine. Depending on the type of engine, the intercooler uses air or water to cool the air as it passes into the cylinders.

Kerosene. The American term for paraffin.

Live hydraulic system. This remains in operation after the transmission clutch has disengaged the drive to the wheels. The hydraulic pump may be mounted on and driven by the engine or in the transmission housing and controlled with an independent or a dual clutch.

Live power take-off. An independent or a dual clutch controls the power take-off, which continues turning after the drive to the transmission has been disengaged.

Lower link sensing. Changes in draft are detected through the lower hydraulic lift arms and relayed to the draft control valve to raise or lower the implement.

Open centre hydraulics. The hydraulic pump provides a constant flow of oil and varies the working pressure according to need.

Orchard tractor. Narrower than a standard model and often equipped with guards to deflect tree branches away from the tractor.

Overhead-valve engine. An engine with the valves located in the cylinder head above the pistons.

PAVT rear wheels. Power-adjusted variable track. Engine power is used to vary the track width after releasing a mechanism locking the centre disc to the rim.

Plow. The American term used to rate a tractor as a one-plow, two-plow, etc until the late 1940s. ('Plow' and 'Bottom' are alternative American terms for furrow.)

Pony engine. The American term for a donkey engine.

Power-assisted steering. A mechanical steering linkage used in conjunction with a hydraulic ram.

Power take-off horsepower. The power available at the power take-off shaft which is less than brake horsepower but greater than that available at the drawbar.

Regular tractor. An American term used for a basic tractor and the same as a standard model in the UK.

Side-valve engine. An engine with the valves at one side of the cylinder block.

Rowcrop tractor. Usually a small tractor with an adjustable track width to suit different row widths. American rowcrop tractors traditionally had a single or twin-vee front wheels. In recent years more powerful tractors have been described as rowcrop tractors.

SAE horsepower. The American standard used for nett engine power.

Styled tractor. Tractors built since the mid-1930s with a sheet metal bonnet, radiator grille and mudguards or wings are described as styled.

Systems tractor. This type of tractor has front and rear implement attachment points and sometimes a load-carrying platform. Most can be operated in either direction and may have root harvesters and other large machines attached around them.

Tool carrier. A three- or four-wheeled frame, normally with the engine and transmission at the rear. Implements may be front-, mid- or rear-mounted on the frame and some tool carriers also have a load-carrying platform.

Tracklaying tractor. A tractor with endless steel or rubber tracks.

Turbocharger. A blower fan driven by the exhaust gases which runs at speeds of 80,000 rpm or more to force extra air into the engine cylinder and increase engine power. The turbocharger compresses the air making it denser and also increases its temperature.

Unitary construction. A tractor assembled by joining the engine transmission and rear axle together without the use of a supporting frame.

Unstyled tractor. Generally made before the 1930s. A tractor without a bonnet or covering over the engine, radiator and steering column was unstyled.

Vision lined. A tractor with the seat and steering wheel offset to provide the driver with a view for rowcrop work.

Wastegate turbocharger. A turbocharger, which runs normally at the lower end of the engine speed range, but at a reduced speed at high engine speeds.

Index

About the Author

Brian Bell MBE

A Norfolk farmer's son, Brian played a key role in developing agricultural education in Suffolk from the 1950s onwards. For many years he was vice-principal of the Otley College of Agriculture and Horticulture having previously headed the agricultural engineering section. He established the annual 'Power in Action' demonstrations in which the latest farm machinery is put through its paces and he campaigned vigorously for improved farm safety, serving for many years on the Suffolk Farm Safety Committee. He is secretary of the Suffolk Farm Machinery Club. In 1993 he retired from Otley College and was created a Member of the Order of the British Empire for his services to agriculture. He is past secretary and chairman of the East Anglian branch of the Institute of Agricultural Engineers.

Brian's writing career began in 1963 with the publication of *Farm Machinery* in Cassell's 'Farm Books' series. In 1979 Farming Press published a new *Farm Machinery*, which is now in its fifth enlarged edition, with 35,000 copies sold. Brian's involvement with videos began in 1995 when he compiled and scripted *Classic Farm Machinery Vol 1*.

Brian Bell writes on machinery past and present for several specialist magazines. He lives in Suffolk with his wife Ivy. They have three sons.

Books and DVDs by Brian Bell

Books in print

Farm Machinery 5th Edition
Machinery for Horticulture (with Stewart Cousins)
Ransomes, Sims and Jefferies
Seventy Years of Farm Machinery: 1. Seedtime
Seventy Years of Farm Machinery: 2. Harvest
Seventy Years of Farm Tractors
Seventy Years of Garden Machinery
The Tractor Ploughing Manual

DVDs

Acres of Change
Classic Combines
Classic Farm Machinery Vol. 1 1940-1970
Classic Farm Machinery Vol. 2 1970-1995
Classic Tractors
Farm Machinery Film Records Vol 1 Grain Grass and Silage
Farm Machinery Film Records Vol 2 Autumn Work and Rootcrops
Farm Machinery Film Records Vol 3 Testing and Prototypes
Harvest from Sickle to Satellite
Ploughs and Ploughing Techniques
Power of the Past
Reversible and Conventional Match Ploughing Skills
Steam at Strumpshaw
Thatcher's Harvest
Tracks Across the Field
Vintage Match Ploughing Skills
Vintage Garden Tractors

Seventy Years of Farm Machinery
Brian Bell

In two volumes Brian Bell deals with the mechanisation of farm machinery in the period 1930–2000. Part 1 looks at the companies and products significant in ploughs and cultivations, drilling, manuring, fertilisers and crop protection. The companion book deals with combines, balers, hay and silage machinery, sugar beet and potato harvesters. The two volumes together are encyclopaedic in scope. Hardback books.

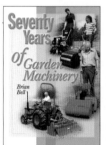

Seventy Years of Garden Machinery
Brian Bell

Covering the period 1920–1990, Brian Bell's book deals with the development of smaller machines: 2-wheeled garden tractors, rotary cultivators, 4-wheeled ride-on tractors, ploughs, drills, cultivators, sprayers, grass-cutting equipment, small trucks, miscellaneous estate items. Hardback book.

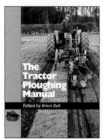

The Tractor Ploughing Manual
Brian Bell

This Society of Ploughmen handbook introduces the types of plough and how to set them up. Then it covers in detail the main classes of competitive tractor ploughing: vintage mounted and trailed; world-type conventional ploughing; and reversible ploughing. Highly illustrated text edited by Brian Bell. Hardback book.

Farm Machinery 5th Edition
Brian Bell

For students and farmers, an outstanding resource book with chapters on tractors, machinery for cultivation, drilling machinery, crop treatment, harvest, maintenance, as well as mechanical handlers, dairy equipment, irrigation, farm power and the workshop. Hardback book.

Farm Machinery Film Records 1, 2 & 3
Brian Bell

This footage from the 1940s and '50s shows a wide range of farm machinery which was in development at the time. It comes from the archive of the National Institute of Agricultural Engineering and its Scottish counterpart. Altogether, over seventy machines are included in the programmes. Experimental prototypes from the NIAE itself are shown. DVDs.

The Farm Tractor Handbook
George Sherwood

The complete facsimile edition of a sought-after book first published in 1918. The author explains the principles of tractor engineering with many examples of tractors then available in the UK. Hardback book.